JOURNAL OF POLYMER SCIENCE: Polymer Symposium   No. 65

Rigid Chain Polymers:
Synthesis and Properties

Proceedings of a
Symposium held during
the March 1977 Meeting of the
American Chemical Society
New Orleans, Louisiana

Editors:

G. C. Berry

C. E. Sroog

Department of Chemistry

Carnegie-Mellon University

Pittsburgh, Pennsylvania 15213

an Interscience® Publication
published by JOHN WILEY & SONS

JOURNAL OF POLYMER SCIENCE: Polymer Symposia

Editors:

H. Mark · C. G. Overberger

Editorial Board:

J. J. Hermans · H. W. Melville · G. Smets

G. C. Berry and C. E. Sroog have been appointed Editors for this Symposium by the Editors of the *Journal of Polymer Science.*

Published by John Wiley & Sons, Inc., this book constitutes a part of the annual subscription of the *Journal of Polymer Science* and as such is supplied without additional charge to the subscribers. Single copies can be purchased from the Subscription Department, John Wiley & Sons.

Subscription price, *Journal of Polymer Science,* Vol. 16, 1978: $485.00. Postage and handling outside U.S.A.: $37.00. Please allow four to six weeks for processing a change of address. Back volumes, microfilm, and microfiche are available for previous years. Request price list from publisher.

Contents

Preface

The initial conception of this Symposium was expressed in its tentative title "Rigid Aromatic Polymers." It quickly became evident that the characterization "aromatic" was too restrictive and the title was accordingly changed to encompass the diversity of the polymers actually discussed.

The symposium included 18 papers of which 16 are published here. Although topics range from polymer synthesis to film evaluation, there emerges a focus on rigid chains as a distinct class. Not only do rigid chains impart distinctive material properties, they also offer numerous challenges to conventional means of physicochemical characterization, and in some respects understanding them will require concepts that lie outside the framework that has been elaborated in the five decades of research on more conventional long-chain molecules. It is our hope that this Symposium will be a stimulus to further developments—both in the search for new materials for exacting applications and in the basic science needed to understand behavior at the molecular level.

G. C. BERRY
C. E. SROOG

evaluate

LIQUID–CRYSTALLINE SOLUTIONS OF POLYHYDRAZIDES AND POLY(AMIDE–HYDRAZIDES) IN SULFURIC ACIDS

P. W. MORGAN*

Pioneering Research Division, Textile Fibers Department, E. I. du Pont de Nemours and Company, Wilmington, Delaware 19898

SYNOPSIS

Para-linked aromatic polyhydrazides and poly(amide–hydrazides) can be dissolved in sulfuric acid of about 100% concentration to form liquid–crystalline solutions, and if the solutions are not heated excessively, they are reasonably stable and can be spun into strong fibers. Similar results can be obtained in fluorosulfonic acid and its admixtures with sulfuric acid. Polyhydrazides and copolyhydrazides based on oxalyl, terephthaloyl, chloroterephthaloyl, and 2,5-pyridinedioyl units and poly(amide–hydrazides) based in part on such intermediates as 4-aminobenzhydrazide and terephthalic dihydrazide were examined for formation of optically anisotropic solutions. Polymers with high proportions of meta- and ortho-linked ring units were highly soluble but did not yield liquid–crystalline solutions.

INTRODUCTION

Aromatic polyhydrazides are well known from the work of Frazer, Wallenberger, and Sweeny [1, 2] as precursors to poly-1,3,4-oxadiazoles. Poly-(amide–hydrazides) have been described in detail by Black, Preston, and coworkers [3, 4] and Culbertson and Murphy [5]. High-tenacity high-modulus fibers have been made from poly(terephthalic hydrazide) [6] and poly(amide–hydrazides) with ordered structures [3, 7]. Throughout these studies there is no indication that any of the polymers have ever yielded liquid–crystalline solutions.

In a recent patent [8] and paper [9] we have shown that liquid–crystalline solutions of polyhydrazides can be formed in a selected group of aqueous organic bases and that the spinning of these solutions leads to high-tenacity fibers. This provides the clue that the hydrazide units when associated with organic base form extended chains in solution and suggests the possibility that other solvents may be found to yield ordered solutions from this class of polymers.

Polyhydrazides are known to be rapidly hydrolyzed when dissolved in ordinary concentrated sulfuric acid [1]. On the other hand, Iwakura and co-workers [10]

---

* Address correspondence to 822 Roslyn Ave., West Chester, PA 19380.

Journal of Polymer Science: Polymer Symposium 65, 1–11 (1978)
© 1978 John Wiley & Sons, Inc.   0360-8905/78/0062-0001$01.00

have reported the synthesis of polyhydrazides from monomethylhydrazine and aromatic diacids in fuming sulfuric acid. Alternatively, N-substituted polyhydrazides can be obtained with alkylating agents and polyoxadiazoles in fuming sulfuric acid [11]. The use of unsubstituted hydrazine and diacids in fuming sulfuric acid leads to the formation of 1,3,4-oxadiazoles [12]. Efimova and coworkers in a recent publication [13] have reported that poly(1,4-phenylene 1,3,4-oxadiazole) as a concentrated solution in sulfuric acid exhibits liquid–crystalline properties.

We have found that liquid–crystalline solutions can be formed from a variety of polyhydrazides and poly(amide–hydrazides) in sulfuric acid of about 100% concentration or in this acid admixed with fluorosulfonic acid. The rate of polymer degradation varies with the chemical structure, some polymers being quite stable. The amount of degradation is greatly reduced by keeping the solutions below about 25°.

RESULTS AND DISCUSSION

Polymer Preparation

The preparation of intermediates and the polyhydrazides used in this study is described in the previous paper [9]. Poly(amide–hydrazides) and some of the intermediates were prepared by the methods reported by Black, Preston, and coworkers [3] and Culbertson and Murphy [5]. One example of polymer preparation is given under "Experimental." The range of compositions examined is shown by the tables and figures.

Polyhydrazide Solutions

For the formation of the most stable solutions and retention of maximum molecular weight without dehydration to the oxadiazole structure, sulfuric acid of about 100% was cooled in an ice-water bath, and dry, finely divided polymer was added with stirring. The acid solutions were essentially colorless when the starting polymer was free of color. The solutions listed in Table I were prepared in this way and examined by polarizing microscope at 27°.

The liquid–crystalline state of the solutions was recognized by the presence of bright opalescence in the bulk solution on stirring and by depolarization of plane-polarized light when a sample was examined by microscope between crossed polarizers. The optically anisotropic solutions were typically highly colored between polarizers and coarsely or finely grained in texture with fluid boundaries to the colored regions. Insoluble material may be birefringent but is recognized by lack of boundary mobility and shape changes on application of pressure to the cover glass.

Table I and Figure 1 provide data for sulfuric acid solutions of one-, two-, three-, and four-component polymers having the hydrazide units, O-2, O-T, O-ClT, O-Pyr, derived from oxalic, terephthalic, chloroterephthalic, and 2,5-pyridinedicarboxylic acids, respectively (see Table I, footnote, for further ex-

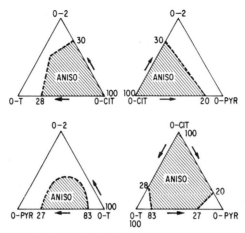

FIG. 1. Regions of observed anisotropic solution formation for polyhydrazides in sulfuric acid.

planation of codes). The figure shows the four equilateral faces of a tetrahedral plot of the observed regions of formation of optically anisotropic solutions, based on an examination of 40 polymers with different composition. The regions might be extended somewhat in some parts by the preparation of polymers with higher molecular weight. However, the principal reason for failure to observe optical anisotropy was insufficient polymer solubility to reach a critical concentration.

O-CIT was the most soluble homopolymer, as it was in quaternary ammonium bases and was the only homopolymer of this group to form an anisotropic solution in sulfuric acid. O-T readily formed anisotropic solutions in aqueous organic bases but dissolved relatively poorly in 100% sulfuric acid and degraded rather rapidly.

The most soluble polymers were the copolyhydrazides. Many O-Pyr copolymers dissolved very rapidly; yet, as already pointed out, the homo O-Pyr did not dissolve well enough to yield an anisotropic solution.

One of the most hydrolytically stable polyhydrazides is (O-2-O)-CIT. Cold concentrated solutions showed little viscosity drop for many hours. In dilute solution at 30° the η_{inh} half-life was about 10 hr (Fig. 2). The initial η_{inh} in dimethyl sulfoxide–5% LiCl was 2.75. The measured initial η_{inh} in 99.7% sulfuric acid was 2.52.

Critical Concentration Points

The usual plot of bulk solution viscosity versus polymer concentration can be obtained but was not attempted because of the need for large samples of polymer. The polymer degradation, even though slow at low temperatures, produces some drift of the transition point with time.

Figure 3 shows a series of concentration points for transition from isotropic to anisotropic state for O-T-O-CIT of various molecular weights. The system does not show the rapid rise in critical concentration at low molecular weights that has frequently been observed in other polymer systems.

TABLE I

Polyhydrazides and Their Solutions in Sulfuric Acid

Polymer composition[a]	η_{inh}[b] (dl/g)	Solution in sulfuric acid		
		Sulfuric acid concentration (%)	Weight % polymer in solution	Solution state
One and two components:				
(O-T-O)-T	4.72	99.9	<10	Isotropic and solid
O-T-O(T/C1T)(90/10)	0.88	99.9	<10	Isotropic and solid
(80/20)	0.39	99.9	<10	Isotropic and solid
(50/50)	0.43	99.9	<10	Isotropic and solid
(O-C1T-O/O-T-O)-T (60/40)	0.92	100.6	19.7	Anisotropic
(70/30)	1.43	100.6	19.7	Anisotropic
(O-T-O)-C1T	1.16	101.0	12	Anisotropic
(O-C1T-O)-T	1.07	100.6	20	Anisotropic
O-C1T-O-(T/C1T)(50/50)	1.21	99.7	19.7	Anisotropic
(O-C1T-O)-C1T	0.77	99.7	19.7	Anisotropic
O-T-O-(T/2, 5Pyr)(90/10)	0.17	99.9	<10	Isotropic and solid
(80/20)	0.64	99.9	<10	Isotropic and solid
(50/50)	0.43	99.7	19.7	Anisotropic
(O-T-O)-2,5Pyr	0.70[d]	99.9	11.0	Anisotropic
(O-T-O/O-2,5Pyr-O)-T (70/30)	0.46	100.6	19.7	Isotropic
(60/40)	0.60	100.6	19.7	Anisotropic
(O-2,5Pyr-O)-T	1.74[c]	99.7	10.9	Anisotropic
(O-T-O/O-2,5Pyr-O)-2,5Pyr (70/30)	0.49	100.6	19.7	Anisotropic
(60/40)	0.53	100.6	19.7	Anisotropic
(50/50)	1.09	99.9	<20	Isotropic and solid
(O-2,5Pyr-O)-2,5Pyr	0.86[d]	99.9	<20	Isotropic and solid
	0.84[c]			
(O-C1T-O)-2,5Pyr	1.43	99.7	18	Anisotropic
	1.98			

(O-C1T-O/O-2,5Pyr-O)-2,5Pyr (50/50)	1.04	100.6	19.7	Anisotropic
(O-2-O)-2	0.31	99.9	<20	Isotropic and solid
(O-2-O/O-T-O)-2 (50/50)	0.49[d]	99.9	<20	Isotropic and solid
(12.5/87.5)	0.38[d]	100.6	<14	Isotropic and solid
(O-2-O)-T	0.24[d]	100.6	<14	Isotropic and solid
(O-2-O/O-C1T-O)-2 (50/50)	0.33	100.6	19.7	Isotropic
(O-2-O)-C1T	2.06	99.9	19.7	Anisotropic
(O-2-O)-2,5Pyr	0.98[d]	99.9	<20	Isotropic and solid
	0.56[c]			
Three components:				
O-2-O-(T/C1T) (90/10)	1.05[d]	99.9	<20	Isotropic and solid
	0.92[c]			
O-2-O-(T/C1T) (80/20)	1.10[d]	99.9	<20	Isotropic and solid
(70/30)	0.35[d]	100.6	19.7	Anisotropic
	0.70[c]			
O-2-O(T/C1T) (60/40)	0.49[c]	100.6	19.7	Anisotropic
	0.57[d]			
(50/50)	0.38[c]	99.7	19.7	Anisotropic
O-2-O-(T/2,5Pyr) (90/10)	0.63[d]	99.9	<20	Isotropic and solid
	1.03[c]			
(75/25)	0.59[d]	99.9	<20	Isotropic and solid
	0.72[c]			
(50/50)	0.48[c]	99.9	<20	Isotropic and solid
(O-2-O/O-2,5Pyr-O)-T (70/30)	1.18	100.6	19.7	Anisotropic
(O-2-O/O-T-O/O-2,5Pyr-O)-T (50/20/30)	0.69	100.6	19.7	Anisotropic
Four components:				
(O-2-O/O-T-O/C1T/2,5Pyr) (25/25/25/25)	0.82	100.6	19.7	Anisotropic

[a] 2, T, C1T, and 2,5Pyr represent oxalyl, terephthaloyl, chloroterephthaloyl, and 2,5-pyridinedioyl units, introduced as diacid chlorides; O-2-O, O-T-O, O-C1T-O, and O-2,5Pyr-O are corresponding dihydrazide units, introduced as the dihydrazides.

[b] Determined in dimethyl sulfoxide–5% LiCl unless otherwise noted.

[c] 5% aqueous diethylamine.

[d] 100% H_2SO_4 at 25°.

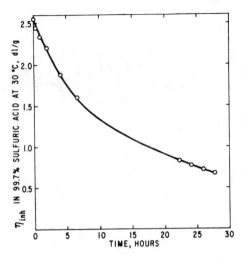

FIG. 2. Inherent viscosity change with time for poly(oxalic–chloroterephthalic hydrazide) in sulfuric acid.

Poly(amide–Hydrazide) Solutions

Para-linked poly(amide–hydrazides) were prepared by use of 4-aminoben-zhydrazide or oxalic and terephthalic dihydrazides as the diamine component (Table II). Since these polymers contain only para-linked phenylene rings, all would be expected to form optically anisotropic solutions if the molecular weight and solubility were high enough. Several, not shown, failed because of insufficient solubility. Although some warming may be used to improve solubility, the loss in molecular weight through degradation may counteract this gain toward attainment of a liquid crystalline solution.

When fluorosulfonic acid is admixed with the sulfuric acid, solubility of the polymers is usually improved. The solutions can be kept cool, and liquid–crys-

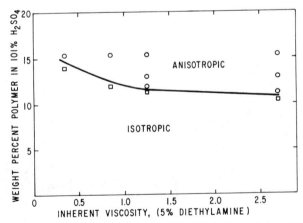

FIG. 3. Critical concentration at 0° versus inherent viscosity for poly(terephthalic–chloro-terephthalic hydrazide).

TABLE II
Poly(amide–Hydrazides) and Their Solutions in Sulfuric Acids

Polymer composition[a]	η_{inh} (dl/g)	Solution in sulfuric acid[f]		
		Sulfuric acid concentration (%)	Weight % Polymer in solution	Solution state
(1,4B-O)-T	2.68[b]	101.0	21	Anisotropic
(1,4B-O)-2,5Pyr	0.99[c]	101.0	10	Anisotropic
(1,4B-O)-BB	0.53[c]	101.0	10	Isotropic and solid
(1,4B-O)-2,5DC1T	0.38[b]	101.0	20	Isotropic and solid
[(O-T-O)-T]-1,4B (50/50)	0.48[d]	101.0	10	Isotropic
			20	Anisotropic
(1,4B-O-T-O-1,4B)-T	1.92[e]	101.0	15	Anisotropic
(1,4B-O-T-O-1,4B)-2,5Pyr	0.23[b]	101.0	10	Isotropic
			20	Anisotropic
(O-T-O/PPD)-T (80/20)	1.8[d]	101.0	10	Isotropic and gel
(70/30)	2.3[d]	101.0	10	Isotropic and gel
(70/30)	2.3	FSO$_3$H/H$_2$SO$_4$ (1/1)	20	Anisotropic
(20/80)	2.18[d]	FSO$_3$H/H$_2$SO$_4$ (1/1)	20	Anisotropic
(O-2-O/PPD)-T (70/30)	1.84[d]	FSO$_3$H/H$_2$SO$_4$ (1/1)	20	Anisotropic
(50/50)	2.24[d]	FSO$_3$H/H$_2$SO$_4$ (1/1)	20	Anisotropic

[a] See Table I for some codes. Additional codes are 1,4B-O, 4-aminobenzhydrazide; 1,4B, 4-aminobenzoyl; BB, 4,4'-bibenzoyl; 2,5DC1T, 2,5-dichloroterephthaloyl; PPD, 1,4-phenylenediamine; 1,4B-O-T-O-1,4B, bis(4-aminobenzoyl) terephthalic dihydrazide.

[b] Dimethylsulfoxide.

[c] Dimethylacetamide-5% LiCl.

[d] 100% sulfuric acid.

[e] Dimethyl sulfoxide-5% LiCl.

[f] Materials mixed at ~0°; examined by microscope at 27°.

[g] Anisotropic at 35–45° for short time.

talline solutions are often obtained from polymers that did not dissolve well enough in 100% sulfuric acid.

Critical concentrations for poly(terephthaloyl 4-aminobenzhydrazide) were determined in 100.2% sulfuric acid at 0° for samples with η_{inh} from 1.8 to 3.8. The solution of polymer with η_{inh} of 1.8 (dimethylacetamide–5% LiCl) was isotropic at 8.8% by weight and optically anisotropic at 10.0%. The higher-molecular-weight polymers formed anisotropic solutions in the same concentration range but dissolved so slowly that degradation made the results irregular.

Nonconforming Polymers

Experience with aromatic polyamide solutions has shown that in order to produce an optically anisotropic solution most of the ring units or other amide-connecting units must have their chain-extending bonds oppositely directed and in coaxial or parallel arrangement. This requirement applies as well to the polyhydrazides and poly(amide–hydrazides). For example, chain units with ortho- or meta-directed links or those based on polymethylene units can be present only in limited proportions without loss of capability to form liquid–crystalline solutions. Table III lists some polymers that did not yield anisotropic solutions at 20% solids or above.

N-Methyl substitution of the hydrazide units in all-parapolyhydrazides (Table III) prevented development of liquid crystallinity in sulfuric acid solutions, just as was the case with aqueous organic bases [9].

The softening temperatures are given for the poly(amide–hydrazides) in Table III to show the depression of melting points produced by substituents and meta and ortho chain extension.

Fiber Preparation

Optically anisotropic spinning dopes of polyhydrazides were prepared in 100.0% sulfuric acid or 1:1 mixtures of sulfuric acid with fluorosulfonic acid. The polymer content was generally 20% by weight. The solutions were prepared by cooling the sulfuric acid to about 10° or the sulfuric–fluorosulfonic acid mixtures to about 2°, after which the polymer was added in small portions with manual stirring at such a rate that the temperature of the mixture did not rise above 40°. After the polymer was in, the mixture was mixed further mechanically while keeping the temperature at 25° to 30°.

Spinning was carried out by the method of Blades [14] by extrusion through 10–20 hole spinnerets into a water bath at 1°. The washed and dried fibers had the properties shown in Table IV.

Fibers from the second spin were passed rapidly through a tube at 285° without extension and showed a 20% gain in tenacity. They then had medium crystallinity and an orientation angle of 13° as measured by wide-angle X-ray diffraction [14b].

TABLE III

Some Isotropic Solutions of Polyhydrazides and Poly(amide–Hydrazides)

| Polymer composition | | η_{inh}^a (dl/g) | Softening temperature (°C)^e | Solution | |
Diamino component	Diacid chloride			Sulfuric acid (%)	Weight % polymer
Isophthalic dihydrazide	Isophthalic	0.76	350	100.0	19.7
	5-tert-Butylisophthalic	0.51^b	300	100.0	21.0
	Terephthalic	1.40	390	100.0	22.0
Terephthalic bis(monomethylhydrazide)	Terephthalic	0.38	340	100.4	19.7
	Chloroterephthalic	0.37	350	100.4	21.5
	2,5-Pyridinedioyl	0.28	>360 dec	100.4	24.0
4-Aminobenzhydrazide	Isophthalic	0.98^c	>380	101.0	21.5
	5-tert-Butylisophthalic	0.49^c	350	101.0	21.5
	5-Chloroisophthalic	0.67^c	340	101.0	21.5
	Sebacyl	0.93^c	275	101.0	21.5
3-Aminobenzhydrazide	Terephthalic	0.43^c	310	101.0	21.5
2-Aminobenzhydrazide	Sebacyl	0.43^d	150	101.0	21.5
	Terephthalic	0.40^d	~385	101.0	21.5

a Dimethyl sulfoxide–5% LiCl except as noted.
b m-Cresol.
c Dimethyl sulfoxide.
d Dimethylacetamide–5% LiCl.
e Point at which a plastic mass could be formed on a gradient temperature bar.

TABLE IV
Fibers from Poly(oxalic–chloroterephthalic Hydrazide)

Polymer η_{inh} (dl/g)	Solution		Fiber properties[a]			
	Solvent	Weight % polymer	Denier	T	E	Mi
2.75	100% H_2SO_4	20	2.69	11.0	5.3	346
2.08	H_2SO_4–FSO_3H	20	1.65	10.2	6.5	326

[a] Tenacity and initial modulus in grams per denier; break elongation in percent.

EXPERIMENTAL

Polymer Preparation

Poly(oxalic–chloroterephthalic Hydrazide)

A suspension of oxalic dihydrazide (17.72 g, 0.15 mole) and 25.2 g LiCl in 600 ml of dimethylacetamide in a 1-liter resin kettle, under nitrogen, was stirred for 0.5 hr and cooled in an ice-water bath. Chloroterephthaloyl chloride (35.62 g, 0.15 mole) was added dropwise through an addition funnel over a period of 72 min with stirring and cooling. The contents of the kettle were allowed to stand overnight and warm to room temperature, after which solid lithium carbonate (11.08 g) was added to neutralize the hydrogen chloride. The clear, viscous solution was combined slowly with a large amount of vigorously stirred water in a blender to precipitate the polymer, which was collected, washed—five times with water, once with 1:1 acetone–water, and once with acetone—and dried under vacuum at 100°. The polyhydrazide had an η_{inh} value of 2.75 (dimethyl sulfoxide–5% LiCl); yield 40.2 g.

Inherent Viscosity Determination

Inherent viscosity [$\eta_{inh} = \ln(\eta_{rel})/c$] was determined at 30° and a polymer concentration c of 0.5 g/100 ml of solution. The traditional solvent was dimethyl sulfoxide with 5% by weight of lithium chloride. However, not all polyhydrazides dissolve in this solvent. More generally useful viscosity solvents are the aqueous organic bases [9]. For this purpose 5% by weight of diethylamine in water was used. Alternatively, 100% sulfuric acid at 25° can be used for polymers with poor solubility. Some poly(amide–hydrazides) form dilute solutions in dimethyl sulfoxide alone or in dimethylacetamide containing lithium chloride.

Fiber preparation and examination of hydrolytic stability were done by Dr. E. H. Barber and Dr. J. D. Hartzler. W. F. Dryden, Jr., provided excellent technical assistance.

REFERENCES

[1] A. H. Frazer and F. T. Wallenberger, *J. Polym. Sci. Part A,2,* 1137, 1147, 1171 (1964).
[2] A. H. Frazer, W. Sweeny, and F. T. Wallenberger, *J. Polym. Sci. Part A,2,* 1157 (1964).
[3] W. B. Black and J. Preston, "High Modulus Wholly Aromatic Fibers," Dekker, New York, 1973; see also *J. Macromol. Sci. Chem., A7*(1), 3–348 (1973).

[4] J. Preston, U.S. Patent 3,632,548 (1/4/72), assigned to the Monsanto Co.

[5] B. M. Culbertson and R. Murphy, *J. Polym. Sci. Part B,4,* 249 (1966); *B,5,* 807 (1967).

[6] A. H. Frazer, U.S. Patent 3,536,631 (10/27/70), assigned to the du Pont Co.

[7] J. Preston, W. B. Black, and W. L. Hofferbert, Jr., *J. Macromol. Sci. Chem., A7*(1), 45, 67 (1973).

[8] J. D. Hartzler, U.S. Patent 3,966,656 (6/29/76), assigned to the du Pont Co.

[9] J. D. Hartzler and P. W. Morgan, "Liquid Crystalline Solutions from Polyhydrazides in Aqueous Organic Bases," presented at the ACS Biennial Polymer Symposium, Key Biscayne, Fla., November 1976. Papers published as "Contemporary Topics in Polymer Science," E. M. Pearce and J. R. Schaefgen, Eds., Plenum, New York, 1977.

[10] Y. Iwakura, K. Uno, S. Hara, and S. Kurosawa, *J. Polym. Sci. Part A-1, 6,* 3371 (1968).

[11] H. Sekiguchi and K. Sadamitsu, U.S. Patents 3,642,708 (2/15/72) and 3,644,297 (2/22/72), assigned to The Furukawa Electric Co.

[12] Y. Iwakura, K. Uno, and S. Hara, *J. Polym. Sci. Part A3,* 45 (1965).

[13] S. G. Efimova, N. P. Okromchedlidze, A. V. Volokhina, and M. M. Iovleva, *Vysokomol. Soedin. Ser. B, 19*(1), 67-69 (1977); Chem. Abstr., *86*(14), 107153 (1977).

[14] H. Blades, (a) U.S. Patent 3,767,756 (10/23/73) and (b) U.S. Patent 3,869,430 (3/4/75), assigned to the du Pont Co.

PREPARATION OF POLYAMIDES VIA THE PHOSPHORYLATION REACTION. I. WHOLLY AROMATIC POLYAMIDES AND POLYAMIDE–HYDRAZIDES

J. PRESTON and W. L. HOFFERBERT, JR.

Monsanto Triangle Park Development Center, Inc., Research Triangle Park, North Carolina 27709

SYNOPSIS

Aromatic polyamides of the AB, AA–BB, and ordered copolyamide types plus aromatic hydrazide and amide–hydrazides were prepared via the phosphorylation reaction. The relationship of the molecular weight, as evidenced by the inherent viscosity values obtained, of the several polymers to the structural features and properties of the monomers used and the polymers obtained is discussed in detail. Some salient observations are that poor solubility of the rodlike polymers definitely appears to set limits to obtaining high molecular weight and strongly basic groups, e.g., —CO—NH—NH$_2$, in monomers, prevent the attainment of high molecular weight because such groups coordinate more strongly with the phosphite complex than does the pyridine catalyst in the phosphorylation reaction. Fibers were spun from two aromatic polyamides prepared via the phosphorylation reaction. Excellent tensile properties were achieved for fiber from poly-p-benzamide (PPB) and the polyisophthalamide of 4,4′-methylenedianiline despite the relatively low molecular weight of these polymers (η_{inh} 1.6 and 1.1, respectively). The PPB fiber was spun from both isotropic and anisotropic spinning solutions. Fiber obtained from the anisotropic dope had moderately high tensile strengths (8–12 gpd) and remarkably high initial modulus values (435–745 gpd). These results indicate that polymers made by the phosphorylation route would give fibers with as high tensile properties as polymer made by the diacid chloride–diamine polycondensation route providing that the inherent viscosities of polymer prepared by the former process could be increased to equal those by the latter process.

INTRODUCTION

Since about 1960, a considerable research effort has been carried out worldwide on preparation and characterization of aromatic polymers for end uses where high heat resistance is required and, more recently, where ultra-high-strength high-modulus fibers are needed. Most of these polymers have been formed by the polycondensation of aromatic diacid chlorides with aromatic diamines to yield aromatic polyamides of the AA–BB type. However, because of the difficulties encountered in the preparation of diacid chlorides of suitable purity, a relatively small number of aromatic diacid chlorides have been used in making a relatively large number of aromatic polyamides. Because of the difficulty of making monomers that contain both an amine group and an acid chloride group in the same monomer, very few aromatic polyamides of the AB type have been prepared.

Recently, an elegant synthesis for aromatic polyamides was reported that

Journal of Polymer Science: Polymer Symposium 65, 13–27 (1978)
0360-8905/78/0062-0013$01.00

involves the direct polycondensation of aromatic amino acids or aromatic diamines with aromatic diacids in the presence of an aryl phosphite and an organic
base. In the original version of this novel synthesis for the preparation of polyamides, first reported by Ogata [1] and referred to as a phosphorylation reaction
[2], rather low-molecular-weight polymer was produced. Two significant
modifications, viz., use of pyridine as the organic base for the reaction [3] and
addition of LiCl to the solvent for the reaction [4] in the process by Yamazaki
led to a considerable increase in the molecular weight of the polymers produced:

$$
HO-CO-Ar-CO-OH \quad \xrightarrow[\text{NMP - 5\% LiCl}]{\left(\bigcirc\!-\!O\right)_3 \ \ P/\text{pyridine}}
$$

$$
\xrightarrow{\ NH_2-Ar'-NH_2\ }
$$

$$
\left[NH-Ar'-NH-CO-Ar-CO\right] + 2 \bigcirc\!-\!OH + 2HO-P\left(O\!-\!\bigcirc\right)_2 \qquad (1)
$$

At present, the phosphorylation reaction is certainly a very valuable means for
screening new polyamides because relatively small amounts of simple monomers,
i.e., diamines, diacids, and aminoacids, are required for synthesis. Although side
reactions in the phosphorylation limit the molecular weight of the products,
polymers having adequate molecular weight for characterization as fibers and
films may be obtained. Undoubtedly, however, commercial fibers will require
polymer of higher molecular weight than that afforded by the phosphorylation
reaction as it is currently practiced.

In this article some work is reported that (1) discloses certain structural
features of selected monomers that make them unsuitable for use in the phosphorylation reaction, (2) gives data on some polymers not previously reported
as having been synthesized by the phosphorylation reaction, and (3) presents
fiber properties for each an AB polyamide and an AA–BB polyamide prepared
via the phosphorylation reaction.

DISCUSSION AND RESULTS

Polymerizations

The significance of Yamazaki's contributions to the phosphorylation reaction lie in (1) the discovery that pyridine appears to have the optimum basicity to form a complex with a diacid and a phosphite ester and (2) the addition of LiCl to amide types of solvents leads to polymer of higher molecular weight. The presence of no base at all or bases weaker than pyridine does not lead to the requisite complex. The presence of a strong base displaces pyridine from the complex. Thus, aliphatic diamines do not yield high-molecular-weight polyamides via the phosphorylation reaction, even when pyridine is present because these strong bases become tied up with the complex in place of pyridine.

The addition of LiCl to amide types of solvents, such as N-methylpyrrolidone (NMP), usually prevents precipitation of the polyamide produced. Thus, the polymer chain does not cease to grow because of phasing out of solution. In general, however, polymers containing a high proportion of para-oriented rings are insoluble or are difficultly soluble in the reaction media. Hence, such polymers obtained via the phosphorylation reaction are generally of low to medium molecular weight. LiCl also appears to activate the solvent in some manner as yet not understood because some polymers that are soluble in NMP without added LiCl are obtained in higher molecular weight when LiCl is present.

Reactions were carried out in NMP containing 5% dissolved LiCl, essentially as reported [4] by Yamazaki except that, in all cases, the temperature of the reaction as conducted by us was that obtained by means of a boiling water bath used to heat the reaction vessel. Numerous experiments in which solvent, base, monomer concentration, etc., were varied convinced us that the conditions specified by Yamazaki had been optimized, except for the temperature of the reaction. Thus, we did not always find that the inherent viscosity of polymer produced from p-aminobenzoic acid varied with temperature in the manner described [4] by Yamazaki. For both convenience and reproducibility of results, a water bath at the boil was selected for all the experiments reported here.

AB polyamides

In Table I are given structures of some AB types of polyamides prepared from various aminobenzoic acids. The inherent viscosity of 1.7 for I (Table I) is in agreement with that previously reported by Yamazaki. This value is high because of the rodlike character of I; the actual molecular weight of I is probably no higher than that of II having an inherent viscosity of about 0.4. It is of interest to note here that the inherent viscosity values of I in organic solvents containing dissolved inorganic salts are considerably higher than those of I in sulfuric acid (Fig. 1).

Despite the para-phenylene rings in III (Table I), this polymer is not of the rodlike type because of the ether linkage that introduces a "kink" into the chain. Thus, the lower inherent viscosity for III relative to I does not necessarily reflect

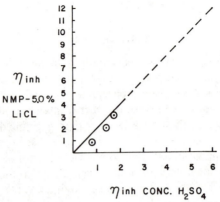

FIG. 1. Relationship between η_{inh} for poly-p-benzamide in N-methylpyrrolidone and in concentrated sulfuric acid. (The departure from linearity at the lower η_{inh} values in sulfuric acid probably is due, at least in part, to degradation in this solvent.)

a lower degree of polymerization. In fact, I, II, and III probably have comparable degrees of polymerization.

The relatively low inherent viscosity obtained for IV (Table I) is somewhat surprising because aliphatic diacids generally react with aromatic diamines in the phosphorylation reaction to give polyamides of higher inherent viscosity than those obtained using aromatic diacids. Separation of the carboxylic acid group from the benzene ring in the monomer used to prepare IV might reasonably have been expected to make it perform more nearly like an aliphatic acid group and hence yield a polymer of high inherent viscosity.

Attempts to prepare V (Table I) in high molecular weight probably failed because of the steric hindrance afforded by the methyl group on the amine function. The inherent viscosity of 0.1 obtained for V is the same as that for the monomer used to make V, suggesting that V is extremely low in molecular weight and is probably only an oligomer (mp 315°–325°; mp monomer, 163°–165°).

AA–BB Polymers

The structures of some polyamides based on isophthalic and terephthalic acids are given in Table II. In the case of those polymers previously reported by Yamazaki, the inherent viscosity values obtained by us are in good agreement with his. Although we made several determinations in dimethylacetamide (DMAc) containing 5% dissolved LiCl and Yamazaki used concentrated sulfuric acid exclusively, the inherent viscosity values obtained for the polymers in Table II are not very different in either solvent. This is in striking contrast to the situation found for rodlike polymers (see previous Discussion and Fig. 1).

The low inherent viscosity obtained by us for 2, a polyterephthalamide, is in good agreement with that reported by Yamazaki [4].

Considering the low inherent viscosities usually obtained for polyterephthalamides, it is surprising to find that polymers 5 and 8 have rather high values, that for 5 approximating that for 1, and that for 8 approximating that for 7.

TABLE 1
AB Polyamides Prepared via the Phosphorylation Reaction

No.	Structure:	η_{inh}^a
I	$\left[-NH-\bigcirc-CO- \right]$	$3.1 \ (1.7)^b$ 1.7^c
II	$\left[NH-\bigcirc-CO \right]$	$0.40 \ (0.43)^b$
III	$\left[-NH-\bigcirc-O-\bigcirc-CO- \right]$	0.5
IV	$\left[-NH-\bigcirc-CH_2-CO- \right]$	0.26
V	$\left[-N-\bigcirc-CO- \right]$ (CH$_3$)	0.1
VI	$\left[-NH-\bigcirc-CO- \right]$ (OH)	0.1
VII	$\left[-NH-\bigcirc-CO- \right]$ (HO)	0.13

a Determined at 25° for a solution of 0.5 g of polymer in 100 ml
of dimethylacetamide containing 5% dissolved lithium chloride.

b Values in parentheses were reported by Yamazaki [4] for solutions
of polymer in concentrated sulfuric acid at 30°.

c Determined at 30° for a solution of 0.5 g of polymer in 100 ml of
concentrated sulfuric acid.

The inherent viscosity obtained for poly-*m*-phenyleneisophthalamide 7 is
surprisingly high, comparable with that for Nomex fiber when it was at an early
stage of commercialization as HT-1, and is probably the highest value reported
to date for a nonrodlike aromatic polyamide prepared via the phosphorylation
reaction.

The inherent viscosity for 9 is about that which would be expected for a
polyisophthalamide having this structure. The low inherent viscosity for poly-
p-phenyleneterephthalamide 10 is not unexpected on the basis of the well-known
low solubility of this rodlike polymer in organic solvents.

Ordered Copolyamides

Some polymer structures based on diamines and diacids with preformed amide
linkages are given in Table III. Because of the regular structure of such polymers
containing a number of monomeric units, these have been referred to as ordered
copolyamides [5]. Typical reactions for their formation are on the following
page.

Ordered copolyamides are generally prepared at low temperatures with diacid
chlorides, and hence no rearrangement of structural units is expected. Because
the phosphorylation reaction does occur at a much higher temperature than that
used with diacid chlorides and, possibly more to the point, under conditions of

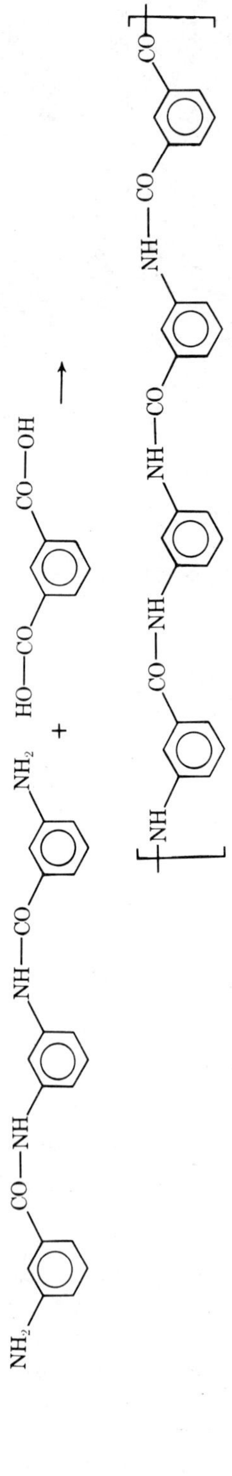

(2)

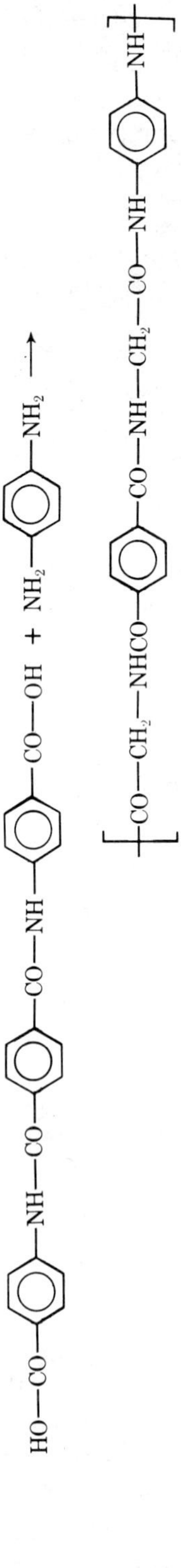

(3)

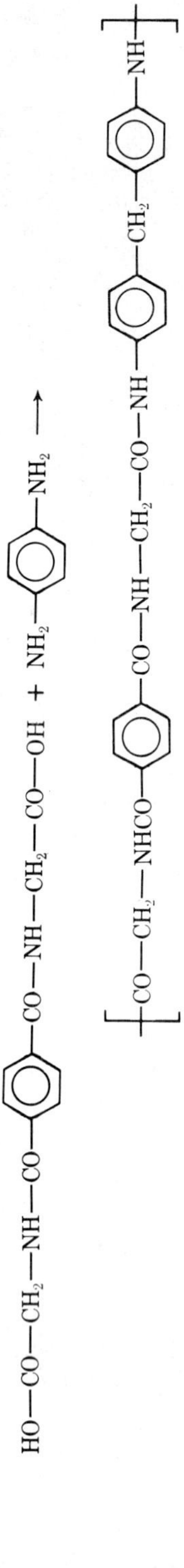

(4)

TABLE II
AA–BB Polyamides Prepared via the Phosphorylation Reaction

No.	Structure	η_{inh}^a DMAc–5% LiCl	η_{inh}^b Conc H_2SO_4
1	[—NH—⟨○⟩—CH₂—⟨○⟩—NH—CO—⟨○⟩—CO—]	1.1(0.93)[c]	
2	[— —CO—⟨○⟩—CO—]	...	0.30(0.33)[c]
3	[— —CO—CH=CH—CO—]	...	1.45
4	[—NH—⟨○⟩CH₂—⟨○⟩—NH—CO—⟨○⟩—CO—]	0.082	
5	[— —CO—⟨○⟩—CO—]	1.0	
6	[—NH—⟨○⟩—C(CH₃)₂—⟨○⟩—NH—CO—⟨○⟩—CO—]	0.97(0.55)[d]	
7	[—NH—⟨○⟩—NH—CO—⟨○⟩—CO—]	1.4	
8	[— —CO—⟨○⟩—CO—]	...	1.37
9	[—NH—⟨○⟩—NH—CO—⟨○⟩—CO—]	0.83	
10	[— —CO—⟨○⟩—CO—]	...	0.28(0.19)[c]
11	[— —CO—CH=CH—CO—]	...	0.24
12	[—NH—⟨○⟩—CH₂—⟨○⟩—NH— —CO—⟨○⟩—P(O)(C₆H₅)—⟨○⟩—CO—]	1.1	

[a] Determined at 25° for a solution of 0.5 g of polymer in 100 ml of solvent.

[b] Determined at 30° for a solution of 0.5 g of polymer in 100 ml of solvent.

[c] Values in parentheses were reported by Yamazaki [4] for solutions of polymer in concentrated sulfuric acid at 30°.

[d] Value in parentheses is for a sample prepared in NMP without added LiCl.

TABLE III

Preparation of Some Ordered Copolyamides via the Phosphorylation Reaction

No.	Structure:	η_{inh} [a]
1'	[NH—⬡—CO-NH—⬡—NH-CO—⬡—NH⋮CO—⬡—CO]	1.3
2'	[⋮CO—⬡—CO]	1.2
3'	[NH—⬡—NH⋮CO—⬡—NH-CO—⬡—CO-NH—⬡—CO]	0.7[b]
4'	[NH—⬡—CH₂—⬡—NH⋮CO—⬡—NH-CO—⬡—CO— —NH—⬡—CO]	0.63
5'	[NH—⬡—CH₂—⬡—NH-CO-CH₂-NH-CO—⬡—CO— -NH-CH₂-CO]	0.9
6'	[NH—⬡—CO-O-CH₂-CH₂-O-CO—⬡—NH-CO—⬡—CO]	0.54

[a] Determined at 25° for 0.5 g of polymer dissolved in 100 ml of dimethylacetamide containing 5% dissolved LiCl.

[b] Determined at 30° for 0.5 g of polymer dissolved in 100 ml of concentrated sulfuric acid.

stepwise amide formation that might be reversible, it was of interest to compare the ordered copolyamide 2' prepared by the phosphorylation reaction and the low-temperature polycondensation route. Oriented films of polymer 2' prepared by the two routes showed no significant differences in differential thermograms. X-Ray diffraction patterns of the films were likewise comparable. The ordered amide–ester 6' does not show a very high inherent viscosity value in view of the fact that this polymer is quite soluble in amide solvents and is made from a diamine having relatively reactive amine groups. Possibly the preformed ester linkage is susceptible to attack by phenol, acid, etc., generated during the course of the reaction.

Polyhydrazides and Polyamide–Hydrazides

The structures of some polymers containing hydrazide and amide plus hydrazide linkages are given in Table IV. For the same reason discussed in connection with the rodlike polyamides, the fact that the inherent viscosity value for A is higher than those for B and C does not necessarily mean a higher molecular weight for A, a rodlike structure, whereas B and C are nonrodlike

TABLE IV
Preparation of Some Polyamide–Hydrazides via the Phosphorylation Reaction

No.	Structure	n_{inh} [a] DMSO
A	$-[-NH-\bigcirc-CO-NH-NH-]-CO-\bigcirc-CO-]-$	0.41
B	$-[-NH-NH-CO-\bigcirc-CO-NH-NH-CO-\bigcirc-CO-]-$	0.23
C	$-[-NH-NH-CO-CO-NH-NH-CO-\bigcirc-CO-]-$	0.35
D	$-[-NH-\bigcirc-CO-NH-NH-]-CO-\bigcirc-CO-]-$	0.70
E	$-[-NH-\bigcirc-CO-NH-NH-CO-\bigcirc-CO-NH-NH-CO-\bigcirc-NH--CO-\bigcirc-CO-]-$	2.43
F	$-[-NH-\bigcirc-CO-NH-NH-CO-\bigcirc-CO-NH-NH-CO-\bigcirc-NH--CO-\bigcirc-CO-]-$	0.73

[a] Determined at 25° for 0.5 g of polymer in 100 ml of solvent.

structures. In fact, it is surprising that the inherent viscosity obtained for A is not considerably higher. Thus, one might reasonably expect that the amine group of the aminobenzyhydrazide monomer should be more reactive in the phosphorylation reaction than the hydrazide groups present in the monomers used for B and C because an amine group is less basic than a hydrazide group and, therefore, less likely to complex with the phosphite intermediate than pyridine (see "Polymerizations" above). When it is considered that polymers A, B, C, and D in Table IV are based on monomers containing hydrazide groups and that strong bases, e.g., aliphatic diamines, complex more strongly with the phosphite complex than does pyridine, it becomes clear why the polymers obtained have low inherent viscosity. It is less clear why polymer D has a higher inherent viscosity value than A. Possibly the answer lies in the fact that D is more soluble than A and achieves a higher molecular weight than A.

It is of interest to note that the inherent viscosity obtained using the Yamazaki process (pyridine catalyst) is comparable with that for B using the original Ogata process (imidazole catalyst). Undoubtedly the high basity of the hydrazide group in the monomer, as discussed above, is responsible for this result.

Use of a diamine containing preformed hydrazide groups dramatically illustrates the points above concerning differences in reactivity for amine and terminal hydrazide groups, which are illustrated on the following page. Thus, reaction of BAB-TDH with isophthalic acid gives E (Table IV) having a high

bis(4-aminobenzoyl)terephthaloyldihydrazide
BAB-TDH

(5)

inherent viscosity value (2.43). Reaction of BAB-TDH with terephthalic acid gives the rodlike polymer F (Table IV) having a low inherent viscosity. The lower solubility of F compared with E may explain the lower molecular weight obtained for F. The higher inherent viscosity of F compared with A can probably be accounted for on the basis of higher reactivity by the amine groups in the monomer used to make F and complexing with the phosphite intermediate of the terminal hydrazide group in the monomer used to make A.

As to the molecular weight of polymer E, it would appear that this polymer is as high in molecular weight as the "partially ordered" polyamide–hydrazide (η_{inh} = 2.32) obtained [6] by reacting 2 moles of of p-aminobenzyhydrazide with 1 mole of terephthaloyl chloride followed by reaction with 1 mole of isophthaloyl chloride. It also appears that polymer E is higher in molecular weight than the "wholly ordered" polyamide–hydrazide prepared [7] by low-temperature polycondensation of BAB-TDH with isophthaloyl chloride. The inherent viscosity of 2.43 probably approaches the limit that one would expect for a polyamide–hydrazide containing three para-oriented benzene rings to one meta-oriented benzene ring. Thus, the phosphorylation route appears, in the case of polymer E, to be an adequate route for making polymer of fiber-forming molecular weight, since fiber of >10 gpd was reported [6] for the "partially ordered" polyamide–hydrazide.

Fiber Obtained from I

It was of interest to spin fibers of I made via the phosphorylation route for comparison with fibers spun from I made by the conventional polycondensation of p-aminobenzoyl chloride hydrochloride. Fortunately Kowlek, [8] in her extensive studies of the polymerization and spinning of I, reported fibers from polymers with an inherent viscosity of only 1.4 (1.7 for the as-spun fibers). Although no attempt was made to optimize the wet-spinning conditions used, remarkably good fibers for such low-molecular-weight polymer (Table V) were obtained by spinning both a low-solids dope (approximately 6% solids) and a relatively high-solids dope (approximately 12%). Contrary to the usual situation for a solution of a given polymer, the solution of lower solids was considerably more viscous than the solution of higher solids. These results would be quite surprising were it not for the fact that lower-solids dopes of I have been shown [9] to be isotropic and the higher-solids dopes have been shown to be anisotropic. (The relationship between viscosity and concentration of poly-p-benzamide with regard to the isotropic and anisotropic states is shown schematically in Fig 2.) Further, anisotropic dopes have been shown [10] to give considerably stronger and stiffer fibers of I than do isotropic dopes (continued on p. 24):

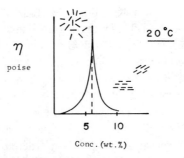

FIG. 2. Schematic representation (after Papkov [9]) of viscosity versus concentration of poly-*p*-benzamide in relation to isotropic and anisotropic states.

Type of spinning dope	$T/E/M_i$
Isotropic	1.2/9.0/64
Anisotropic	7.2/8.1/283

The fiber obtained by us from the spinning of the isotropic dope was considerably better than that reported by Kwolek for the spinning of an isotropic dope. However, the dope used by Kwolek was obtained from a single dope separated into isotropic and anisotropic phases. Nevertheless, our data do show a considerable difference between fiber obtained from spinning a low-solids dope of I versus spinning a high-solids dope of I.

In Table VI, data on the fiber obtained from the high-solids dope (about 12%) of I are compared with fiber of I (spun from about 10% solids dope) reported [8] by Kwolek, both fibers having been prepared from polymers having com-

TABLE V

Fiber Tensile Properties[a] from Poly-*p*-benzamide of Inherent Viscosity 1.6[b]

	$T/E/M_i$ [c]	W-T-B	Denier
Spun from 12% solids dope			
As-spun (air dried)	8.0/4.7/435 (8.7/5.4/439)	0.251 (0.304)	4.2 (4.6)
Drawn (by hand) at 275°C	12.3/3.0/588 (14.5/3.6/620)	0.228 (0.314)	3.6 (3.9)
Redrawn (by hand) at 400°C	11.3/2.1/745 (12.7/2.3/707)	0.152 (0.176)	3.5 (3.2)
Spun from 6% solids dope			
As-spun (air dired)	3.9/6.7/216 (4.2/8.6/185)	0.222 (0.289)	7.8 (7.8)

[a] Tensile properties are the average in 10 breaks; the highest break in each series is given in parentheses.

[b] Determined at 30° for a solution of 0.5 g of polymer in 100 ml of concentrated sulfuric acid.

[c] $T/E/M_i$ = ten., gpd; elongation to break, %; initial modulus, gpd.

TABLE VI
Tensile Properties[a] of Hot-Drawn Fibers from Poly-p-benzamide

	Wet spun fiber from 12% solids dope[b]		Properties reported by Kwolek for dry spun fiber from 10.2% solids dope[c]	
	$T/E/M_i$[d]	Den.	$T/E/M_i$	Den.
As-spun				
Air dried	8.0/4.7/435	4.2	8.2/3.1/509	3.1
	(8.7/5.4/439)	(4.6)		
Dried on 165°C roller	10.4/4.1/483	3.9	-- -- --	--
	(11.1/4.5/520)	(4.0)		
Heat-treated				
Approx. 300°C	12.3/3.0/588	3.6	10.3/2.4/604	3.0
	(14.5/3.6/620)	(3.9)		
400°C	11.3/2.1/745	3.5	13.2/2.5/671	2.8
	(12.7/3.3/598)	(4.2)		

[a] Tensile properties are the average of 10 breaks; the highest break in each series is given in parentheses.

[b] Solvent NMP containing 5% dissolved LiCl.

[c] Solvent tetramethylurea containing dissolved LiCl.

[d] $T/E/M_i$ = ten., gpd; elongation to break, %; initial modulus, gpd.

parable inherent viscosities. When fibers having comparable thermal histories are compared, it can be seen that comparable results are achieved despite the fact that we used a wet-spinning method and Kwolek used dry spinning. In view of the fact that Kwolek and Blades [11] both have reported ultrahigh-strength high-modulus fibers (to 19 gpd for yarn) from I of high molecular weight (η_{inh} ≈4) prepared by the acid chloride–amine route, it would thus appear from our results that I prepared via the phosphorylation reaction could be expected to yield fibers of comparable properties provided that means to higher-molecular-weight polymer can be found.

Fiber Obtained from 1

Fibers spun from 1 by wet spinning had remarkably good tensile properties

TABLE VII
Fiber Properties of MDA-I

η_{inh}[a]	$T/E/M_i$[b]
1.10[b]	3.6/31/48[c]
2.51[d]	4.1/20/—[d]

[a] Determined at 30° for 0.5 g of polymer in 100 ml of concentrated sulfuric acid.

[b] $T/E/M_i$ = ten., gpd; elongation to break, %; initial modulus, gpd.

[c] Yarn properties; contained numerous fused filaments, possibly a consequence of low molecular weight or residues of by-products from the phosphorylation reaction.

[d] Data from reference 12.

(Table VII) in view of the low inherent viscosity of the sample. These properties are compared (Table VII) with data taken from the literature [12] for a wet-spun sample of 1 having a high inherent viscosity. Both fibers are seen to have roughly comparable tensile strengths when their elongations to break are taken into account. This is not to say that other fiber properties, e.g., fatigue resistance, would not be superior for the higher-molecular-weight sample. Nevertheless, the fiber from 1 prepared via the phosphorylation route certainly appears, in this preliminary evaluation, to have adequate initial fiber properties, and the fiber should be comparable with that from polymer prepared by the more conventional low-temperature polycondensation route provided somewhat higher molecular weight can be achieved.

EXPERIMENTAL

Monomers

The amino acids (Table VIII) used, with the exception of 4'-aminophenyl-4-oxybenzoic acid (mp 286°–290°), were commercial products and were used as received. Diacids were commercial products (Table VIII) and were used as received with the exception of the phosphorus-bridged diacid (mp 335°–350°), which was synthesized by Dr. K. Moedritzer, of the Monsanto Company. The two diacids containing preformed amide linkages, N,N'-bis-p-carboxyphenyl)-terephthalamide (mp >400°) and N,N'-bis(carboxymethyl)terephthalamide (mp 240°–245°) were synthesized by the Schotten-Bauman reaction of terephthaloyl chloride with, respectively, p-aminobenzoic acid and glycine. Purification of these diacids was effected by dissolving them in dilute acid and reprecipitating with hydrochloric acid.

The following simple diamines were commercial products and were used as received: m-phenylenediamine (du Pont), mp 62°–64°, and p-phenylenediamine (Fisher Scientific), mp 138°–140°. The 4,4'-methylenedianiline (Allied

TABLE VIII
Amino Acid and Diacid Monomers

Monomers:	M.P.°C	Supplier:
Amino-acids		
p-Aminobenzoic	185-187	Fisher Sci. Co.
m-Aminobenzoic	167-169	Eastman Org. Chem.
p-Aminophenylacetic	201 (dec.)	Aldrich Chem. Co.
N-Methyl-p-aminobenzoic	155-160	Eastman Org. Chem.
4-Aminosalicylic	147 (dec.)	Gallard-Schlesinger
3.Amino-4-hydroxybenzoic	198-200 (dec.)	Gallard-Schlesinger
Diacids		
Terephthalic	404-408 (subl.)	Amoco Chem.
Isophthalic	344-347	Eastman Org. Chem.
Fumaric	293-295	Eastman Org. Chem.

Chemical) monomer, mp 91°–92°, was distilled, recrystallized, and redistilled prior to use; the 3,3'-methylenedianiline, mp 83°–84°, was obtained from the Burdick-Jackson Company and sublimed. Sublimed 4,4'-isopropylidenedianiline (mp 133°–134°) was prepared by Dr. J. K. Lawson, Jr., of the Monsanto Company.

The diamines containing preformed amide (mp 214°–215°) and hydrazide (mp 360°–363°) linkages have been described elsewhere [5, 6]. The diamine ethylene-bis(p-aminobenzoate), mp 222°–223°, was prepared by reduction of the corresponding dinitro compound. The monomers p-aminobenzhydrazide (Gallard-Schlesinger, mp 225°–227°, and oxalyl dihydrazide (Olin Mathieson), mp 242°–245°, were commercial products, and the isophthaloyl dihydrazide, mp 223°–226°, was prepared by addition of hydrazine to methyl isophthalate.

Polymerizations

The experimental conditions specified by Yamazaki [4] were used with the exception that the temperature was set by a boiling water bath surrounding the reaction vessel.

We wish to thank E. W. Folk for the spinning of the fibers and Dr. K. Koedritzer and Dr. J. K. Lawson, Jr., for the generous quantities of monomers. We are also indebted to Dr. W. B. Black for reviewing the manuscript and making numerous helpful suggestions.

REFERENCES

[1] N. Ogata and H. Tanaka, *Polym. J., 2,* 672 (1971).
[2] N. Ogata and G. Suzuki, "Macromolecular Syntheses" Vol. 5, E. L. Wittbecker, Ed., Wiley, New York, 1974, p. 107.
[3] N. Yamazaki, F. Higashi, and J. Kawabata, *J. Polym. Sci. Polym. Chem. Ed., 12,* 2149 (1974).
[4] N. Yamazaki, M. Matsumoto, and F. Higashi, *J. Polym. Sci. Polym. Chem. Ed., 13,* 1373 (1975).
[5] J. Preston, *J. Polym. Sci. Part A-1, 4,* 529 (1966).
[6] J. Preston, W. B. Black, and W. L. Hofferbert, Jr., *J. Macromol. Sci. Chem., A7*(1), 67 (1973).
[7] J. Preston, W. B. Black, and W. L. Hofferbert, Jr., *J. Macromol. Sci. Chem., A7*(1), 45 (1973).
[8] S. L. Kwolek, U.S. Patent 3,888,965 (1975), assigned to E. I. du Pont de Nemours and Co.
[9] S. P. Papkov, V. G. Kulichikhin, V. D. Kalmykova, and A. Ya. Malkin, *J. Polym. Sci. Polym. Phys. Ed., 12,* 1753 (1974).
[10] S. L. Kwolek, U.S. Patent 3,671,542 (1972), assigned to E. I. du Pont de Nemours and Co.
[11] H. Blades, U.S. Patent 3,767,756 (1973), assigned to E. I. du Pont de Nemours and Co.
[12] H. W. Hill, Jr., S. L. Kwolek, and W. Sweeny, U.S. Patent 3,094,511 (1963), assigned to E. I. du Pont de Nemours and Co.

SYNTHESIS, CHARACTERIZATION, RHEOLOGICAL, AND FIBER FORMATION STUDIES OF p-LINKED AROMATIC POLYAMIDES

HIROSHI AOKI,* DAVID R. COFFIN,† TONY A. HANCOCK,†† DANIEL HARWOOD, RUDOLPH S. LENK,§ JOHN F. FELLERS, and JAMES L. WHITE

Polymer Engineering, The University of Tennessee, Knoxville, Tennessee 37916

SYNOPSIS

Aromatic polyamides are being extensively studied in our laboratories along the lines of synthesis, dilute solution viscosity and light scattering, concentrated solution optical and rheological measurements, and wet spinning of fibers. Several series of polymers have been synthesized, including those starting with chloro- and methyl-substituted p-phenylene diamine and also those starting with the monomer bisacid A_2. Poly(chloro-p-phenylene terephthalamide) in dilute solution shows a marked hydrodynamic volume dependence on solvent when studied in DMA–5% LiCl versus H_2SO_4. At higher concentration, optical characterization shows that liquid crystal formation in the quiescent state corresponds to changes observed in the rheological behavior for Kevlar–H_2SO_4 solutions; these observations include a yield stress and the inversion of the order of $G''/G' = \tan \delta$. Fibers produced by wet spinning and subsequently studied in cross section via SEM have revealed that material continuity is easily attained with Kevlar–H_2SO_4 spinning dopes under a variety of processing conditions.

INTRODUCTION

Para-linked aromatic polyamides have become increasingly important in recent years because of the ability to produce high-modulus and tensile-strength fibers from them [1-5]. These materials are prepared by low-temperature solution and interfacial polycondensation [1, 2, 6-8]. Because of the rigidity of the polymer backbone, they have unusual solution properties [1, 5, 9-11], including the formation of polymer liquid crystals in concentrated solution. According to the classical patent of Kwolek [1], this liquid–crystalline behavior is key to the creation of high-modulus and tensile-strength fibers.

During the past few years, we have been engaged in an extensive program on the synthesis, characterization, solution and rheological properties, and wet

* Present address: Unitika, Ltd., Uji, Kyoto, Japan.
† Present address: Celanese Fibers Company, Charlotte, North Carolina.
†† Present address: IBM, Lexington, Kentucky.
§ Present address: Polytechnic of the South Bank, London, England.

Journal of Polymer Science: Polymer Symposium 65, 29–40 (1978)
© 1978 John Wiley & Sons, Inc.   0360-8905/78/0062-0029$01.00

spinning of aromatic polyamide fibers. Our earlier papers have treated the synthesis of new p-linked [12] and m-linked [13] polyamides and also some block copolymers [14, 15]. Reports have also been prepared on solution and rheological properties [16, 17] and wet spinning [4, 12, 13] for both p- and m-linked polymers. The p-linked polymers that have been investigated in this program include

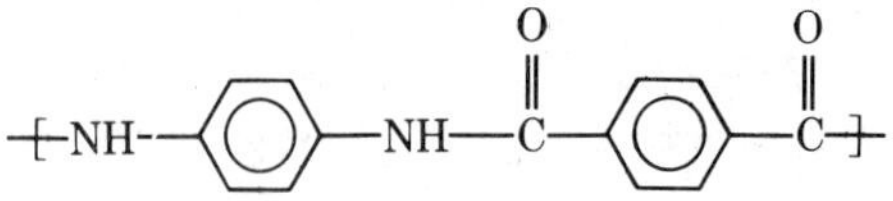

(I)

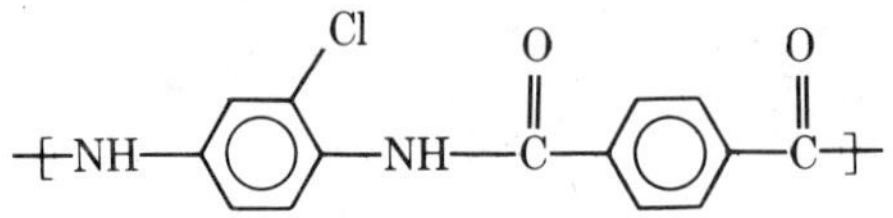

(II)

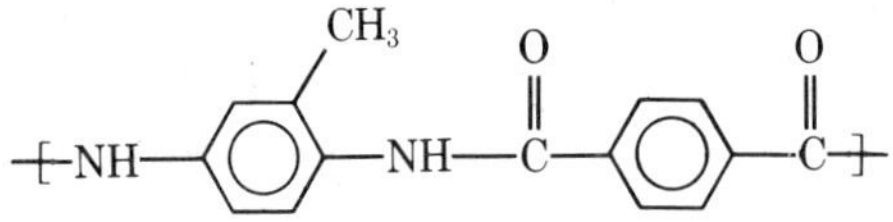

(III)

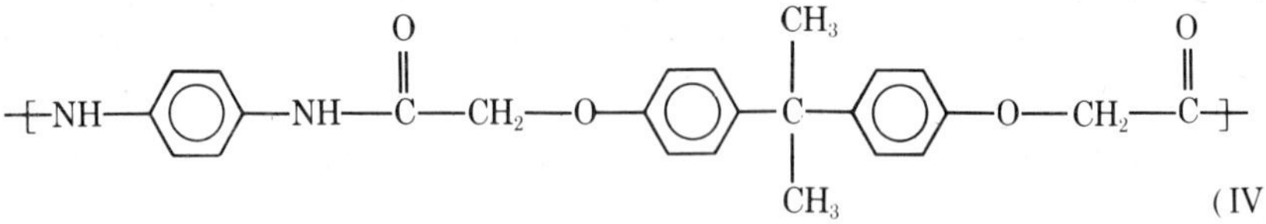

(IV)

In the present article, we critically review our earlier researches and present some new results. Our intention here is to present a broad perspective rather than a detailed view of particular problems.

SYNTHETIC EFFORT

Synthesis has centered on producing molecular structural variations and then determining their effect on solution and mechanical properties. We have used the Shotten-Bauman reaction as applied to diamines and diacid chlorides:

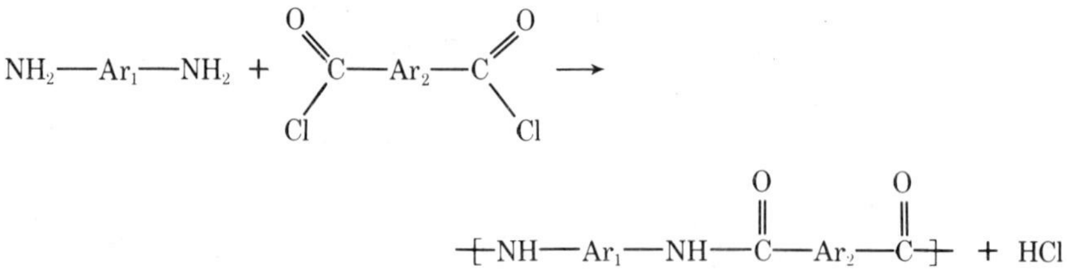

Both low-temperature solution polycondensation [3, 6–8] and interfacial polycondensation processes have been used.

Structures such as IV were synthesized to examine the characteristics of the family of aromatic polyamides based on bisacid A2 and to see how they compare to polymers based on terephthalic acid.

Structures I to III have been synthesized so that sets of these structures covering as broad a range of molecular weights as possible are available. They are being used to investigate both substituent and molecular-weight effects on solution behavior, including optical and light-scattering phenomena.

DILUTE SOLUTION STUDIES

Significant interest exists in developing a fundamental understanding of the hydrodynamic volume, frictional, and conformational characteristics of aromatic polyamides [10, 18–22]. Techniques used in our own studies are light scattering with a Brice-Phoenix Model 2000 and dilute solution viscometry. Solvent systems include sulfuric acid of variable SO_3 percentage and dimethylacetamide (DMA) with small percentages of LiCl. Light-scattering data gathered at various angles and concentrations were treated by the Zimm method [23–26] (Fig. 1). Dilute

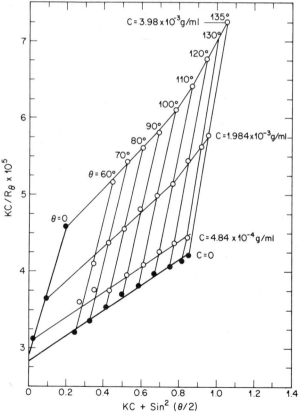

FIG. 1. Zimm plot for poly(chloro-*p*-phenylene terephthalamide) in 96% H_2SO_4, $\overline{M}_w = 3.5 \times 10^4$ g/mole.

solution viscosity data were obtained for studying both the effect of LiCl in DMA and polymer intrinsic viscosities.

These rather standard experimental methods as used on classically randomly coiled polymers are fraught with complications when applied to p-linked aromatic polyamides with their limited solubility. Only solvents that may act as multicomponent systems dissolve many of the p-linked polymers. First of all, sulfuric acid may be viewed as a multicomponent solvent system, i.e., H_2O, SO_3, and H_2SO_4. One questions to what extent each of these components or component combinations interacts with the polymer. One can extend this question to the solvent system DMA–LiCl. Figure 2 shows how varying amounts of LiCl influence the nature of the solvent system. By treating LiCl as a solute and extrapolating to zero concentration, it gives an intrinsic viscosity of 0.2 dl/g. Thus it is reasonable to conclude that such solvent systems have considerable structure. So the dual consideration arises that a multicomponent solvent system can have variable interaction with the polymer and additionally have a structure that is perturbed by the presence of a polymer molecule dissolved in it. Various models have been proposed [27, 28] of solutions of flexible polymer chains with binary solvents that divide into "sea" and "polymer coating" phases.

Note in Figure 3 that the same polymer sample of poly(chloro-p-phenylene terephthalamide) has strikingly different intrinsic viscosities in two different solvent systems. This observation raises the question of whether the polymer has a different effective hydrodynamic volume because of a change (1) in shape of the polymer chain, (2) in the nature of a complex solvent with varying polymer concentration, and possibly (3) polymer association being responsible for the changed intrinsic viscosities.

Further consideration of Figure 1, especially the limiting slope of the lines at constant concentration, may be indicative of an unusual effect. Note that from zero concentration to the highest concentration used, the slope is increasing! Now the slope at zero concentration is customarily used to evaluate the radius of gyration. The increasing value of this slope with increasing polymer concentration is another indication of complex behavior.

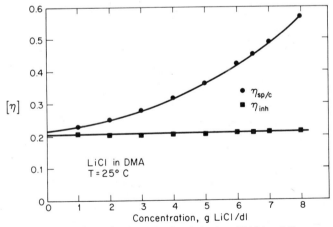

FIG. 2. Effect of LiCl on the flow behavior of DMA at 25°.

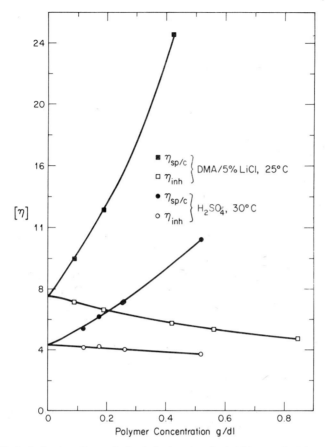

FIG. 3. Effect of solvent quality on the intrinsic viscosity of poly(chloro-*p*-phenylene terephthal-amide).

When the light scattering and intrinsic viscosity data for poly(chloro-*p*-phenylene terephthalamide) in the molecular weight range of 9000–43,000 g/mole dissolved in 96% sulfuric acid at 25°C were used to obtain the Mark-Houwink equation, some rather striking results were obtained. More specifically, corrections for depolarization and fluorescence made a large change in "K" and "a" values. For uncorrected light scattering molecular weights,

$$[\eta] = 3.6 \times 10^{-6} \, \overline{M}_w^{1.3}$$

however when corrections for depolarization and fluorescence were made

$$[\eta] = 9.9 \times 10^{-4} \, \overline{M}_w^{0.8}$$

While the "a" = 1.3 is in reasonable agreement with some other published reports (10, 18, 20), we feel that it is incorrect due to erroneous light scattering molecular weight determinations. Corrections for depolarization and fluorescence must be made (26).

This unusually low "a" value for such a polymer structure may be rationalized in the following manner. The high intrinsic viscosities at low $\overline{M}_w$ make an expanded random coil conformer an unlikely explanation. Rather it is likely that the polymer has a conformer modelled by the wormlike chain concept.

CONCENTRATED SOLUTION STUDIES

Optical and rheological studies have been carried out on the solutions of several polyamides. Greatest attention has been given to investigations of the properties of dissolved DuPont Kevlar® fibers in 100% H_2SO_4. Kevlar® appears to have structure I. Polymers with structure II were also investigated as well as an m-linked aromatic polyamide (redissolved DuPont Nomex® fiber) and various aliphatic polyamides, including poly-γ-benzyl-L-glutamate (PγBLG). The details are given in a report by Aoki, Fellers, and White [16].

First the amount of polarized light passing through a specified thickness of polymer solution in a microscope with crossed polarizer and analyzer (Kwolek DDA test [1]) was determined. The experimental results for several systems are shown in Figure 4. No light passes at low concentrations for any of the polymer solutions studied. However, at high concentrations, there is significant light transmission for the polymers with structures I and II and for the PγBLG, but not for polymers with structure IV or m-linked aromatic or aliphatic polyamides. The measured light transmission indicates these three polymer solutions are birefringent in a state of rest. This means they are in a mesomorphic state

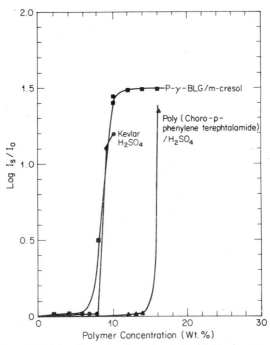

FIG. 4. Plot of reduced light intensity passing through polarized light microscope in Kwolek DDA test as a function of concentration for p-linked aromatic polyamides and P$_\gamma$BLG.

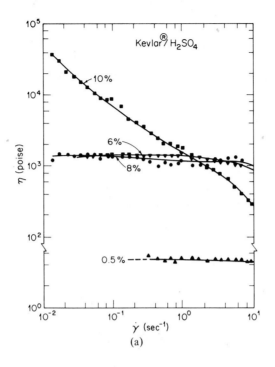

(a)

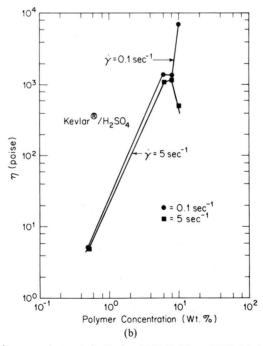

(b)

FIG. 5. (a) Viscosity versus shear rate for Kevlar–100% H_2SO_4 at 25°C; (b) viscosity versus concentration for Kevlar–100% H_2SO_4 at 0.1 sec^{-1} and 5 sec^{-1} at 25°.

or are liquid–crystalline. The existence of concentrated solutions of macro-molecules with structures such as I and II in such a state has been pointed out by Kwolek and her coworkers [1, 11, 29]. The observation of concentrated PγBLG and similar polypeptide solutions in the liquid–crystalline state was made many years ago by Robinson [30–32]. The general area of polymer liquid–crystalline behavior has been reviewed by White and Fellers [5].

The rheological measurements were carried out in a Weissenberg rheogoniometer. Viscosity η and principal normal stress difference N_1 measurements were made in steady shear flow. Dynamic storage modulus $G'(\omega)$ and loss modulus $G''(\omega)$ were determined in oscillatory shear flow. Special precautions

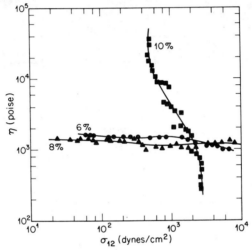

FIG. 6. Viscosity versus shear stress for Kevlar–100% H_2SO_4 at 25°.

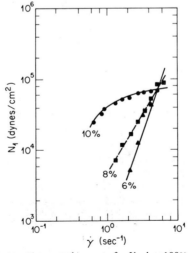

FIG. 7. Normal stresses versus shear rate for Kevlar–100% H_2SO_4 at 25°.

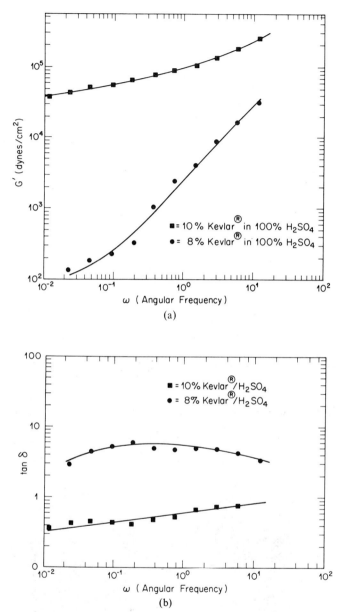

FIG. 8. Dynamic properties of Kevlar–100% H_2SO_4 at 8% and 10% Kevlar. (a) Storage modulus (G') of Kevlar–H_2SO_4 solution at 25°; (b) loss tangent (tan δ) of Kevlar–H_2SO_4 solution at 25°.

were necessary for all solutions of *p*-linked aromatic polyamides where 100% H_2SO_4 was used. These involved a specially designed cone and plate system made of stainless steel and the use of a paraffin oil layer around the solutions to prevent absorption of moisture from the air. Viscosity–shear rate data for Kevlar® solutions at various concentrations at 25° are shown in Figures 5 and 6. Figure

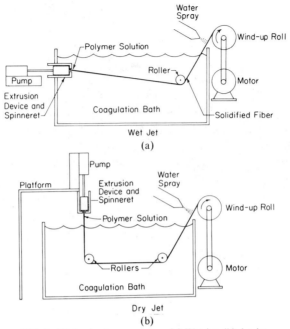

FIG. 9. Wet-spinning apparatus. (a) Wet jet; (b) dry jet.

5 shows at low $\dot{\gamma}$ the anisotropic solution curve has a 45° slope. Replotting against σ_{12}, the data may be seen to represent a yield value as shown in Figure 6. The shear stress during flow never goes below a critical value. These data closely resemble the behavior found by Papkov et al. [9] in poly-p-benzamide solutions. It is to be noted from Figures 5 and 6 that viscosity varies strikingly with concentration. At lower concentrations, viscosity at first increases, then goes through a maximum, and then decreases. Notice, however, that this occurs only at the

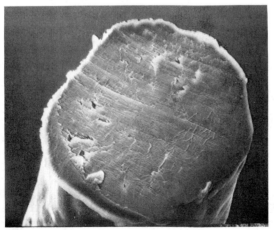

FIG. 10. SEM photomicrograph of wet-spun redissolved Kevlar fiber (500X). From 12% Kevlar in H_2SO_4, coagulant is 15% H_2SO_4 in water.

higher shear rates. Similar behavior has been reported by Kwolek [1, 7] and Papkov et al. [9] for *p*-linked aromatic polyamide solutions and by Hermans [33] for PγBLG polypeptide solutions.

Normal stresses in Kevlar–H_2SO_4 solutions measured at 25° are shown as a function of shear rate for several concentrations in Figure 7. The values are large and increase with increasing concentration levels, especially at lower $\dot{\gamma}$.

Note also in Figure 8 that the energy storage and loss characteristics as measured by tan δ or G''/G' change in a dramatic way from the isotropic (8%) to anisotropic (10%) solutions. The storage modulus increases by greater than a factor of 10, and G' becomes greater than G'' for the 10% solution. These observations compel one to think of these liquid–crystalline dopes as having some "solidlike" characteristics.

Wet Spinning

Fibers of both structures I and IV have been wet-spun using the wet-jet procedure and those of structure I using the dry-jet method (see Fig. 9) [4, 12]. The wet jet is the classical wet-spinning procedure used for rayon and acrylic fiber. The dry-jet method, as shown by Blades [2], is of special utility for *p*-linked aromatic polyamides where it leads to greatly enhanced modulus and tensile strength. Polymer solutions formed from structure I are liquid–crystalline, and fibers wet-spun from it are crystalline and oriented [1–5].

Redissolved Kevlar® solutions in 100% sulfuric acid were dry-jet wet-spun. Spinning-dope solids levels ranged from 1, 2, and 5% in isotropic solutions to 12% in anisotropic or liquid crystalline solutions. The anisotropic dopes were spun into coagulation baths of 5% NaOH in water, water, 5, 10, and 15% sulfuric acid in water. All the dopes showed good spinnability even at the 1% and 2% solids levels. The superstructure of the fiber spun from the anisotropic dope was not changed significantly with coagulation bath concentration. The external surface of the fiber was much rougher at high bath pH than at low pH. The void fraction for the fibers spun ranged from 0.05 to 0.20 and the tenacity from 0.3 to 2.7 g/d as spun. On hot drawing a tenacity of 7.1 g/d was achieved.

Scanning electron microscope photomicrograph (Fig. 10) shows a typical cross section of our dry-jet wet-spun Kevlar®. The fiber is very solid when compared with many other wet-spun fibers such as Nomex®, nylon-6, or nylon-66 [4].

This research has received partial support from the National Science Foundation under Grant DMR 75-21764.

REFERENCES

[1] S. L. Kwolek, U.S. Patent 3,671,542 (1972).
[2] H. Blades, U.S. Patent 3,767,756 (1973).
[3] P. W. Morgan, *ACS Polym. Prepr.*, *17*(1), 47 (1976); and *Macromol.*, *10*, 1381 (1977).
[4] T. A. Hancock, J. E. Spruiell, and J. L. White, *J. Appl. Polym. Sci.*, *21*, 1227 (1977).
[5] J. L. White and J. F. Fellers in "Fiber Structure and Properties," J. L. White, Ed., *Appl. Polym. Symp.*, in press.
[6] P. W. Morgan and S. L. Kwolek, *J. Polym. Sci.*, *40*, 299 (1959).

[7]   P. W. Morgan and S. L. Kwolek, *J. Polym. Sci., 2,* 181 (1964); also with W. R. Sorensen, U.S. Patent 3,063,977 (1962).

[8]   E. L. Wittbecker and P. W. Morgan, *J. Polym. Sci., 40,* 289 (1959).

[9]   S. P. Papkov, V. G. Kulichikin, V. P. Kalmykova, and Y. Y. Malkin, *J. Polym. Sci. Polym. Phys. Ed., 12,* 1753 (1974).

[10] J. R. Schaefgen, V. S. Foldi, F. M. Logulio, V. H. Good, L. W. Gulrich, and F. L. Killiam, *ACS Polym. Prepr., 17*(1), 69 (1976).

[11] M. Panar and L. F. Beste, *ACS Polym. Prepr., 17*(1), 65 (1976); *Macromol., 10,* 1401 (1977).

[12] R. S. Lenk, J. L. White, and J. F. Fellers, *J. Appl. Polym. Sci., 21,* 1543 (1977).

[13] R. S. Lenk, J. F. Fellers, and J. L. White, *Polym. J., (9),* 9 (1977).

[14] R. J. Zdrahala, E. M. Firer, and J. F. Fellers, *J. Polym. Sci., Polym. Chem. Ed., 15,* 689 (1977).

[15] J. F. Fellers, Y. Lee, and R. J. Zdrahala, *J. Polym. Sci., Polym. Sym., 60,* 59 (1977).

[16] H. Aoki, J. F. Fellers, and J. L. White, "An Investigation of the Rheological and Optical Properties of Aliphatic and Aromatic Polyamides," University of Tennessee Polymer Science and Engineering Report No. 107, August 1977, also *J. Appl. Polym. Sci.,* to appear.

[17] H. Aoki, D. Harwood, Y. Lee, J. F. Fellers, and J. L. White, "Solution and Rheological Properties of Poly(*m*-phenylene Isophthalamide) in Dimethylacetamide/LiCl," University of Tennessee Polymer Science and Engineering Report No. 95, August 1977, also *J. Appl. Polym. Sci.,* to appear.

[18] M. Arpin and C. Strazielle, *C. R. Acad. Sci. Ser., C280,* 1293 (1975).

[19] M. Arpin and C. Strazielle, *Makromol. Chem., 177,* 293 (1976).

[20] M. Arpin and C. Strazielle, *Makromol. Chem., 177,* 581 (1976).

[21] M. Arpin, F. Debeauvais, and C. Strazielle, *Makromol. Chem., 177,* 585 (1976).

[22] D. R. Coffin, Ph.D. Dissertation, University of Tennessee, in progress.

[23] B. H. Zimm, *J. Chem. Phys., 16*(12), 1093 (1948).

[24] B. H. Zimm, *J. Chem. Phys., 16*(12), 1099 (1948).

[25] C. Tanford, "Physical Chemistry of Macromolecules," Wiley, New York, 1961.

[26] J. M. Evans, "Light Scattering from Polymer Solutions," M. B. Hughlin, Ed., Academic, New York, 1972, pp. 89–146.

[27] J. Pouchly, A. Ziving, and K. Solc, *J. Polym. Sci. Part C, 23,* 245 (1968).

[28] M. Yamamoto and J. L. White, *Macromolecules, 5,* 58 (1972); and with D. L. MacLean, *Polymer, 12,* 290 (1971).

[29] S. L. Kwolek, P. W. Morgan, J. R. Schaefgen, and L. W. Gulrich, ACS *Polym. Prepr., 17*(1), 53 (1976); and *Macromol., 10,* 1390 (1970).

[30] C. Robinson, *Trans. Faraday Soc., 52,* 571 (1956).

[31] C. Robinson, *Tetrahedron, 13,* 219 (1961).

[32] C. Robinson, *Mol. Cryst., 1,* 467 (1966).

[33] J. Hermans, *J. Colloid Sci., 17,* 638 (1962).

POLYQUINOLINES: A CLASS OF RIGID-CHAIN POLYMERS

W. H. BEEVER and J. K. STILLE*

Department of Chemistry, University of Iowa, Iowa City, Iowa 52242

SYNOPSIS

The reaction product of *m*-cresol and phosphorus pentoxide has been shown by ^{31}P-nmr to consist of an equimolar mixture of mono- and di-*m*-cresyl phosphate. A mixture of pure di-*m*-cresyl phosphate and *m*-cresol is an excellent polymerization medium for the preparation of polyquinolines via the Friedländer reaction, giving higher-molecular-weight polymers than can be obtained from the crude reaction of *m*-cresol and phosphorus pentoxide. Spin-lattice relaxation times T_1 have been used to relate the chain mobility of the polymers in solution to the glass transition temperature (T_g) in the solid.

INTRODUCTION

Thermally stable polymers containing quinoline units in the main chain can be synthesized by a polymerization reaction that allows a variety of structural modifications, resulting in a range of chain stiffness that can be altered from a relatively flexible polymer with a low glass transition temperature to a rodlike molecule with a high glass transition temperature [1, 2]. Most of the materials have high crystalline transition temperatures, but a low degree of crystallinity [3, 4]. For example, polyquinolines 1 containing some flexibility, exhibit lower transition temperatures (T_g = 265–350, T_m = 450–480) and lower crystallinity (<20%) than the rigid polyquinoline 2 (T_g = 415, T_m = 552, >50% crystallinity).

1a R = H
b R = C_6H_5

2

The largely amorphous polyqinolines, of which 1 is representative, are char-

---

* To whom correspondence should be sent. Present address: Department of Chemistry, Colorado State University, Fort Collins, CO 80523.

Journal of Polymer Science: Polymer Symposium 65, 41–53 (1978)
© 1978 John Wiley & Sons, Inc.   0360-8905/78/0062-0041$01.00

acterized by high modulus below the glass transition temperature ($E' = 6 \times 10^{10}$ dynes/cm^2, **1a** Ar = 4,4'-$C_6H_4OC_6H_4$) and good solubility in common organic solvents. In addition, all polyquinolines show excellent thermal stability as determined by thermal gravimetric analysis.

EXPERIMENTAL

Mono- and Di-*m*-cresyl Phosphorochloridates [5, 6]

A solution of 1491 g (890 ml, 9.72 moles) of freshly distilled phosphorus oxychloride and 2102 g (2034 ml, 19.44 moles) of freshly distilled *m*-cresol was heated slowly in the dark to 260° under a static nitrogen atomsphere. This temperature was maintained for 8 days, and the resulting dark yellow solution was fractionally distilled under reduced pressure. A total of 129 g (4%) of mono-*m*-cresyl phosphorochloridate (bp 58° at 0.02 mm) was obtained along with 2225 g (77%) of di-*m*-cresyl phosphorochloridate (bp 113° at 0.02 mm, Lit [5]: bp 228–230° at 22 mm). H^1-nmr (CCl$_4$) mono-*m*-cresyl phosphorochloridate: δ 2.25 (s, 3, CH$_3$), δ 6.8–7.3 (m, 4, aromatic). ^{31}P-nmr (*m*-cresol): δ −6.2 relative to H$_3$PO$_4$. H^1-nmr (CCl$_4$) di-*m*-cresyl phosphorochloridate: δ 2.25 (s, 6, CH$_3$), δ 6.8–7.3 (m, 8, aromatic). ^{31}P-nmr (*m*-cresol): δ −13.9 relative to H$_3$PO$_4$.

m-Cresyl Phosphate (3)

A total of 30 g (0.144 mole) of mono-*m*-cresyl phosphorochloride was cooled in ice, and 5.5 ml of water was slowly added with stirring. The reaction was kept in ice for an additional 1 hr and was then vigorously stirred at room temperature for another 10 hr. The hydrochloric acid was removed *in vacuo,* and the resultant violet oil was placed under reduced pressure for an additional 8 hr, giving 19.8 g (73%) of a clear violet oil. H^1-nmr (CCl$_4$): δ 2.15 (s, 3, CH$_3$), δ 6.7–7.2 (m, 4, aromatic), δ 12.2 (s, 2, OH). ^{31}P-nmr (1.0 g *m*-cresyl phosphate in 2.4 ml *m*-cresol): δ 5.4 relative to H$_3$PO$_4$.

Di-*m*-cresyl Phosphate (4)

A mixture of 500 g (1.69 moles) of di-*m*-cresyl phosphorochloridate and 500 ml of water was stirred for 30 min at room temperature and was then heated to 80°. This temperature was maintained for 1 hr, and then the reaction was cooled, with stirring, to room temperature. The resulting two layers were separated, and the viscous layer was first made weakly basic with 6 *M* sodium hydroxide and then acidified with 6 *N* sulfuric acid. The viscous organic layer was separated and dissolved in 1.1 liters of benzene. This solution was washed five times with 500 ml of water and dried by azeotroping the water, and then the benzene was removed under reduced pressure. The resultant straw-colored oil was placed under reduced pressure in a round-bottomed flask while rotating it in an oil bath at 100°. A total of 357.2 g (76%) of di-*m*-cresyl phosphate (**4**) as a light brown viscuous oil was obtained. H^1-nmr (CCl_4): δ 2.2 (s, 6, CH_3), δ 6.6–7.2 (m, 8, aromatic), δ 13.1 (s, 1, OH); ^{31}P-nmr (2.2 g of di-*m*-cresyl phosphate in 1.1 ml of *m*-cresol): δ 12.3 relative to H_3PO_4.

Reaction of *m*-Cresol and Phosphorus Pentoxide

A suspension of 17 g of P_2O_5 and 50 ml of freshly distilled *m*-cresol was stirred and heated at 140°–145° for 2 hr under a nitrogen atmosphere and then allowed to cool to room temperature. A straw-colored viscuous oil resulted. ^{31}P-nmr: δ 5.3 and δ 11.9 relative to H_3PO_4. Both peaks were of equal intensity. A very small amount of tricresyl phosphate was also obtained at δ 18.5.

Polymerization Using Di-*m*-cresyl Phosphate (4) as Catalyst

A mixture of 0.7629 g (3.000 mmoles) of 4,4′-oxy-bis(acetophenone) (5), 1.2254 g (3.000 mmoles) of 4,4′-oxy-bis(2-benzoyl-benzenamine) (6), 6.94 g (64.2 mmoles) of freshly distilled *m*-cresol, and 20.0 g (72.0 mmoles) of 4 was stirred under a static nitrogen atmosphere at 135°–137° for 12 hr. The resulting dark brown viscuous solution was slowly poured into a stirring solution of 360 ml of ethanol and 30 ml of triethylamine, giving a white fibrous material that was then collected by filtration and extracted for 12 hr with 95% ethanol containing a small amount of added triethylamine. The polymer was air-dried and then dissolved in a minimum amount of tetrachloroethane and reprecipitated into a stirring solution of ethanol. The precipitated polymer was collected by filtration, air-dried, and further dried under reduced pressure at 90° for 18 hr.

Polymerization Using *m*-Cresyl Phosphate (3) as Catalyst

The polymerization using 3 as a catalyst was carried out by a procedure identical with that described for 4, except that chloroform instead of tetrachloroethane was the solvent in the reprecipitation of the polymer. Both polymerizations were run in the same oil bath for exactly the same time. Intrinsic viscosities of both polymers are shown in Table I.

TABLE I
Effect of Polymerization Medium on Viscosity of **7**

Polymerization medium	Intrinsic viscosity (dl/g)	Viscosity solvent
m-Cresyl phosphate	0.462	$Cl_2CH_2CH_2Cl_2$
Di-*m*-cresyl phosphate	0.813	$Cl_2CH_2CH_2Cl_2$
m-Cresyl phosphate	0.532	*m*-Cresol[a]
Di-*m*-cresyl phosphate	1.850	*m*-Cresol[a]

[a] *m*-Cresol was freshly distilled.

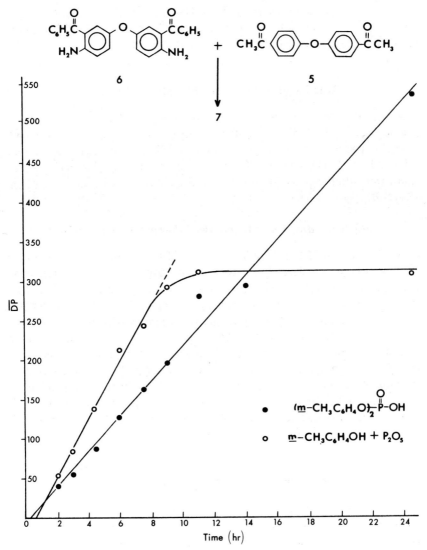

FIG. 1. Kinetics of polymerization of **5** with **6** to give **7**.

Kinetics of the Polymerization of 5 with 6

A comparative rate study of the polymerization of **5** with **6** was carried out as follows. In one resin flask, a solution of 42.5 g (299 mmoles) of phosphorous pentoxide and 100 ml of freshly distilled *m*-cresol was stirred under a static nitrogen atmosphere while being heated to 130° [1, 2]. This temperature was maintained for 2 hr, and then the solution was allowed to cool to room temperature. In a second resin flask, 166 g (598 mmoles) of diester (**4**) was mixed with 56 ml of freshly distilled *m*-cresol. The monomers **5** and **6** were added simulta-

TABLE II
Data on the Polymerization of **5** with **6** in m-Cresol-P_2O_5

Sample No.	Polymerization Time, h	$[\eta]$ [a]	$\overline{Mn}$ [b]	$\overline{DP}$ [c]
1	2	0.54	15 700	54
2	3	0.72	24 700	84
3	4.5	1.01	41 000	138
4	6	1.34	63 000	214
5	7.5	1.46	72 100	244
6	9	1.65	86 200	292
7	11	1.72	92 500	312
8	24.5	1.70	90 800	308

[a] Viscosities determined in chloroform at 25°.
[b] Calculated from Mark-Houwink equation [2]: $[\eta] = (9.0 \times 10^{-4}) \overline{M_n}^{0.66}$.
[c] Calculated from $\overline{M_n}$.

neously to both reaction flasks at room temperature, and the flasks were immersed in an oil bath at 135°. Aliquots of approximately 10 ml were withdrawn at various intervals from each of the reaction flasks and precipitated into ethanol–triethylamine and then reprecipitated from chloroform into 95% ethanol. Each sample was dried at 100°–120° under reduced pressure. The intrinsic viscosity of a chloroform solution of each sample was then determined. The reaction time, intrinsic viscosity, $\overline{M_n}$ and $\overline{DP}$ for each reaction are shown in Tables II and III.

TABLE III
Data for the Polymerization of **5** with **6** in m-Cresol-Di-m-cresyl Phosphate

Sample No.	Polymerization time, h	$[\eta]$ [a]	$\overline{Mn}$ [b]	$\overline{DP}$ [c]
1	2	0.44	11 900	40
2	3	0.54	16 100	54
3	4.5	0.75	26 100	88
4	6	0.95	37 500	126
5	7.5	1.12	48 300	165
6	9	1.27	58 200	196
7	11	1.61	83 500	282
8	14	1.65	86 600	294
9	24.5	2.47	159 000	538

[a] Viscosities determined in chloroform at 25°.
[b] Calculated from Mark-Houwink equation [2]: $[\eta] = (9.0 \times 10^{-4}) \overline{M_n}^{0.66}$.
[c] Calculated from $\overline{M_n}$.

Determination of Spin Lattice Relaxation Times T_1 on Polymers 7 and 8

The broad-band decoupled ^{13}C-nmr spectrum of polymers **7** and **8** were obtained on an approximately 10% (w/w) solution of each polymer in deuterochloroform. The spectra were run on a Bruker HX-90E CW spectrometer at room temperature. The measurement of spin-lattice relaxation times T_1 was carried out by the partially relaxed Fourier transform (PRFT) method [7, 8]. The experiment consisted of time-averaging the pulse sequence, which may be written as $180°\text{-}\tau\text{-}90°\text{-}T$, where T, the delay time between sequences, was long compared with T_1. The interval τ between the $180°$ and $90°$ pulses was varied over a range of values from $\tau \ll T_1$ to $\tau \gg T_1$. The values of T_1 were determined by using an iterative computer program to give the best fit of the observed intensities to the equation:

$$I_i(\tau) = I_{i\infty}[1 - A \exp(-\tau/T_1)]$$

where under ideal conditions $A = 2$ for a $180° - \tau - 90°$ pulse sequence [8, 9].

TABLE IV

Carbon-13 Spectra[a] and Spin-Lattice Relaxation Times[a] for Polyquinoline **7**

7

Carbon No.	Obsd (Calcd)	T_1 (sec)	Carbon No.	Obsd (Calcd)	T_1 (sec)
1	158.0 (164.6)	3.39 ± 0.67	9	129.1 (128.9)	0.35 ± 0.08
2	155.0 (158.7)	3.95 ± 1.02	10	128.9 (128.4)	0.21 ± 0.04
3	154.4 (158.4)	3.10 ± 0.59	11	128.5 (127.3) (127.4)	0.28 ± 0.05
4	148.2 (149.7)	2.93 ± 0.67	12	126.3 (128.9)	3.55 ± 0.94
5	145.8 (149.1)	3.64 ± 1.05	13	122.7 (119.5)	0.15 ± 0.02
6	137.9 (141.6)	4.34 ± 1.02	14	119.3 (115.5)	0.11 ± 0.06
7	134.7 (133.9)	3.49 ± 0.81	15	119.1 (114.5)	0.18 ± 0.04
8	131.9 (131.3)	0.16 ± 0.02	16	112.3 (114.1)	0.17 ± 0.02

[a] Spectra obtained in deuterochloroform at room temperature.

TABLE V
Carbon-13 Spectra[a] and Spin-Lattice Relaxation Times[b] of Polyquinoline **8**

8

Carbon No.	Obsd (Calcd)	T_1 (sec)	Carbon No.	Obsd (Calcd)	T_1 (sec)
1	157.1 (163.3)	4.70 ± 1.16	12	122.2 (115.3)	0.13 ± 0.01
2	156.4 (158.7)	3.05 ± 0.49	13	118.0 (114.5)	0.16 ± 0.01
3	154.4 (158.2)	2.49 ± 0.31	14	113.4 (113.9)	0.12 ± 0.01
4	146.8 (148.4)	5.47 ± 1.19	15	131.3 (131.1)	b
5	144.2 (141.6)	2.84 ± 0.54	16	127.4 (127.8)	b
6	138.1 (141.6)	3.05 ± 0.44	17	127.3 (127.4)	b
7	136.4 (140.9)	2.79 ± 0.41	18	131.3 (128.9)	b
8	136.0 (133.9)	3.28 ± 0.58	19	127.7 (127.4)	b
9	133.2 (132.5)	6.49 ± 1.65	20	131.3 (128.9)	b
10	129.9 (128.4)	0.18 ± 0.01	21	132.4 (127.3)	b
11	126.3 (127.3)	0.14 ± 0.01			

[a] Spectra obtained in deuterochloroform at ambient temperature.
[b] These peaks were not resolved well enough to obtain T_1.

RESULTS AND DISCUSSION

Polymerization

The polymerization reaction of an aromatic bis-*o*-aminoketone with an aromatic bisketomethylene monomer has been shown [1, 2] to give high-molecular-weight polyquinolines by acid catalysis. Optimum polymerization rates and molecular weights were obtained in mixtures of either *m*-cresol and polyphosphoric acid or *m*-cresol and the reaction product of *m*-cresol and phosphorus pentoxide [1, 2]. The later reaction product, as expected, consisted of an equimolar mixture of mono- (**3**) and di-*m*-cresyl phosphate (**4**), as the stoichiometry suggests:

$P_2O_5 + 3m\text{-}CH_3\text{—}C_6H_4OH \longrightarrow$

$$(m\text{-}CH_3C_6H_4O)\text{—}\overset{\displaystyle O}{\overset{\|}{P}}\text{—}OH \quad + \quad (m\text{-}CH_3C_6H_4O)_2\ \overset{\displaystyle O}{\overset{\|}{P}}OH$$

$$\underset{\displaystyle OH}{}$$

                    3   **4**

Both esters **3** and **4** were independently synthesized [5, 6] and purified, and their ^{31}P-nmr spectra were compared with the ^{31}P-nmr spectrum of the reaction product of heating *m*-cresol with phosphorus pentoxide. The slight differences in chemical shifts can be attributed to concentration differences in the samples run.

The effectiveness of **3** or **4** as a polymerization medium was evaluated using each product separately with *m*-cresol in a polymerization of **5** with **6**. The results (Table I) showed that although a mixture of *m*-cresol and ester **3** was marginally effective as a polymerization medium, a mixture of di-*m*-cresyl phosphate and *m*-cresol proved to be very effective. The kinetics of the polymerization of **5** with **6** in both *m*-cresol–P_2O_5 and *m*-cresol–di-*m*-cresyl phosphate are second-order, as shown by a plot of $\overline{DP}$ as a function of time (Fig. 1). Although the polymerization in the *m*-cresol–phosphorus pentoxide mixture slows down and virtually stops after 12 hr at a $\overline{DP}$ of 320, the polymerization in *m*-cresol and di-*m*-cresyl phosphate remains linear even after 24 hr, achieving a $\overline{DP}$ of 538. Thus, one important advantage of the independently synthesized diester **4** is that much higher molecular weights can be obtained. The upper $\overline{DP}$ in *m*-cresol–phosphorous pentoxide possibly is a result of side reaction with one or more of the monomer functional groups. Acid-catalyzed cyclotrimerization has been suggested [2] as a reaction that not only would consume ketomethylene groups but also would lead to branching.

Polymer Properties

Linear rigid-chain macromolecules generally exhibit greater mechanical strength and higher phase-transition temperatures than the more flexible-chain polymers. Most polymers with highly rigid recurring units in the chain, however, are either crystalline, or have high melting (softening) temperatures and are insoluble in suitable solvents as a result of ring (particularly aromatic ring), spiro, or ladder structures in the main chain. The good solubility of polyquinolines in common organic solvents not only allows an opportunity to study the solution properties but also provides a means of fabrication. Polymers that owe their rigidity to a high percentage of aromatic units offer unique opportunities for studying the relationship between molecular motion of chains and their effect on the thermal and solution properties. In the absence of appreciable intramolecular chain attraction, such as a hydrogen bonding, the internal rotations of chain segments are greatly restricted and can be correlated with phase transitions, relaxations, and conformation in solution.

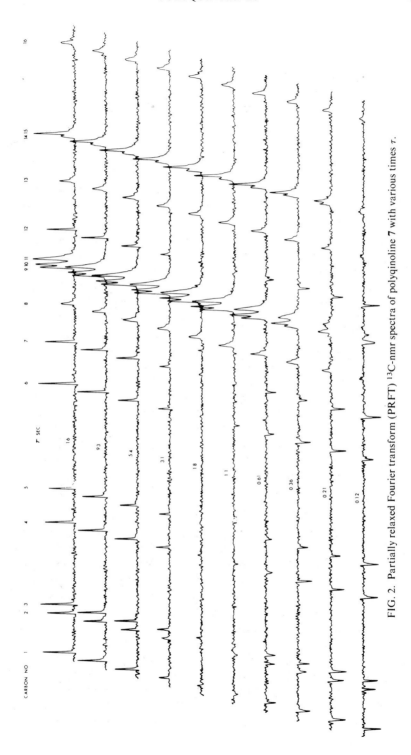

FIG. 2. Partially relaxed Fourier transform (PRFT) ^{13}C-nmr spectra of polyquinoline 7 with various times τ.

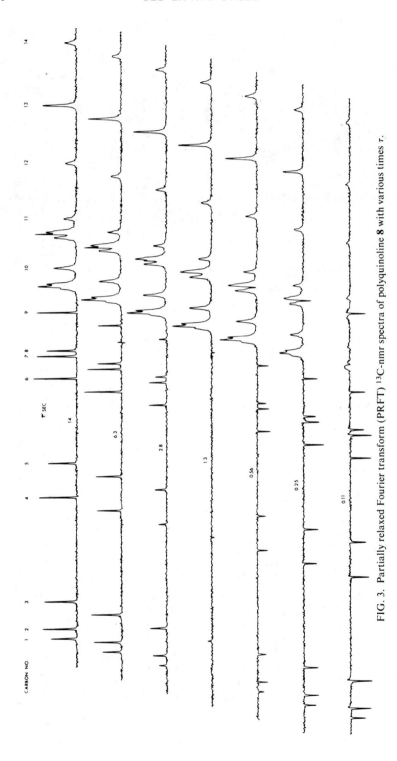

FIG. 3. Partially relaxed Fourier transform (PRFT) ^{13}C-nmr spectra of polyquinoline 8 with various times τ.

TABLE VI
Properties of Unoriented Film Sample of Polyquinoline 7

Thickness (mil)	0.3–0.4	
η_{inh} (0.5% CHCl$_3$ 30°)	1.59 (M_n = 81,700)	
Density (g/cc)	1.174	
X-Ray	Amorphous	
T_g (°C)	266	
Tensile	23°	200°
Modulus (psi)	1.1 × 10^6	0.74 × 10^6
Elongation (%)	0.6	2.0
Tensile strength (psi)	3.2 × 10^3	3.5 × 10^3

Electrical	
Dielectric strength (kv/mil)	7.8
Volume resistivity (ohm-cm/mil)	2.5 × 10^{15}
Dielectric constant	2.6/2.5 (10^2/10^{5}Hz)
Dissipation factor	0.017/0.005 (10^2/10^{5}Hz)

Thermal stability	
TGA (5% wt loss)	525° (air)
	630° (N$_2$)
Film failure (air aging)	2 hr at 450°
	20 hr at 350°
	11 days at 300°
	6 months at 250°
	>10 years at 200°*

* Extrapolated from Arrhenius type of plot.

Permeability	

O$_2$ (cc/100 in^2-24 hr-atm per mil) 278
Water vapor (g/100 in^2-24 hr-44 mm per mil) 9.4

The difference between the glass transition temperatures (δT_g = 42 ± 3°) of a series of the more heavily substituted polyquinolines (1b) and those in which the 3-phenyl substitution is absent (1a) was attributed [5] to restricted rotation in the chain. In particular, substitution of phenyl for hydrogen should hinder rotation of the 2-phenylene unit.

^{13}C spin-lattice relaxation mechanisms are generally dominated by dipole-dipole interaction with neighboring nuclei, modulated by molecular motion. Because carbon atoms a and, especially, b in 7 and 8 have identical near neighbors, spin-lattice relaxation times T_1 can be related to molecular motion and thus to chain mobility in solution.

The broad-band decoupled ^{13}C-nmr spectra for 7 and 8 are recorded in Tables IV and V.

The calculations for carbons in the quinoline ring were made assuming the chemical shifts for quinoline [9] to be base values and then adding the influence of phenyl substituents [10]. Similarly, the calculated values for carbons in phenyl rings were determined by assuming the value of 128.5 ppm for benzene and then adding substituent effects [10].

PRFT spectra observed for the carbons of polymers **7** and **8** are shown in

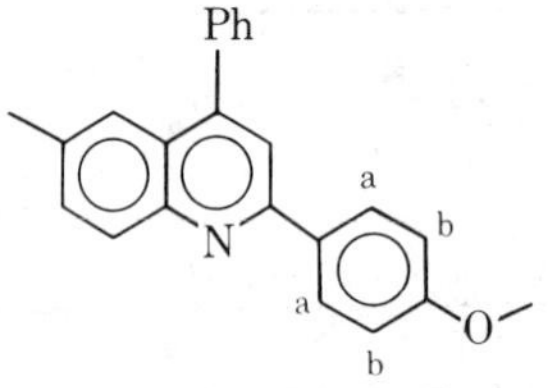

7 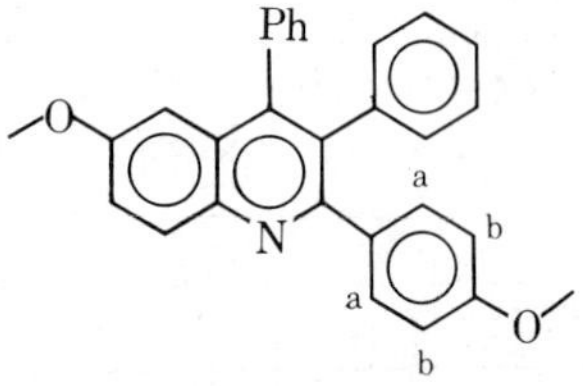   **8**

Figures 2 and 3. The T_1 values, which were determined by an iterative computer program, are listed in Tables IV and V.

The relatively fast relaxation times for all carbons is characteristic of high polymers in solution. Dipole-dipole relaxation is most efficient when the motion frequency compares with the resonance frequency. Thus a correlation time t_c of approximately 10^{-9} sec leads to the most efficient relaxation. This value can be as low as 10^{-12} to 10^{-13} for small molecules, but for highly restricted polymers it may approach 10^{-9}. Any change that shortens t_c lengthens T_1; thus large molecules relax more efficiently than small, tumbling molecules. A comparison of carbons 10 or carbons 15 and 13 of polymers **7** and **8**, respectively, shows in both cases that the T_1 values are longer for polymer **7**, corresponding to a shorter t_c. This implies that polymer **7** is relaxing more efficiently through greater molecular motion in solution. The difference in the motion of the carbons in question in these two polymers can be attributed to restricted rotation of the 2-phenylene group as a result of the 3-phenyl substitution. This restricted rotation should correspond to a higher T_g for polymer **8** in the solid, which is in agreement with the values [5] observed (**7**, $T_g = 266°$; **8**, $T_g = 305°$).

In a preliminary run, polyquinoline **7** was spun from an NMP dope containing 12% solids at 150° into concurrent nitrogen at 240°. The relatively low percentage of elongation did not allow draw ratios greater than about 2, and the drawn fiber, unfortunately, did not show appreciable crystallinity. Tenacity and moduli of 3 and 45 gpd, respectively, were obtained. Unoriented film cast from chloroform has high modulus, modest tensile strength, and low elongation combined with good dielectric properties (Table VI). Oxygen and water vapor permeabilities are relatively high.

The thermal stability of **7** is excellent. A life in air of about 10 years at 200° can be obtained by extrapolation of the Arrhenius data for isothermal aging. These times represent film failure and are exceptionally good. This excellent stability is not surprising, since quinoline itself has been reported [11] to have the highest decomposition temperature of any of the aromatic rings.

This work was supported by grants from the Army Research Office, Research Triangle, N.C., and the National Science Foundation. We wish to thank Dr. G. Pearson, University of Iowa, for the T_1 measurements, Dr. R. Angelo, du Pont, for the data on film samples of polyquinolines, and Dr. J. J. Kleinshuster and Dr. R. S. Irwin, du Pont, for spinning fiber and providing tensile testing.

REFERENCES

[1] J. F. Wolfe and J. K. Stille, *Macromolecules, 9,* 489 (1976).
[2] S. O. Norris and J. K. Stille, *Macromolecules, 9,* 496 (1976).
[3] W. Wrasidlo and J. K. Stille, *Macromolecules, 9,* 505 (1976).
[4] W. Wrasidlo, S. O. Norris, J. F. Wolfe, T. Katto, and J. K. Stille, *Macromolecules, 9,* 512 (1976).
[5] K. A. Petrov, E. E. Nifantev, and R. F. Nikitina, *J. Gen. Chem. USSR, 31,* 1592 (1961).
[6] J. R. Ferraro and D. F. Peppard, *J. Phys. Chem., 67,* 2639 (1963).
[7] R. Freeman and H. D. W. Hill, *J. Chem. Phys., 53,* 4103 (1970).
[8] V. D. Mochel, *J. Macromol. Sci. Rev. Macromol. Chem., 8,* 289 (1972).
[9] R. J. Pugmire, D. M. Grant, M. J. Robins, and R. K. Robins, *J. Am. Chem. Soc., 91,* 6381 (1969).
[10] G. C. Levy and G. L. Nelson, "Carbon-13 Nuclear Magnetic Resonance for Organic Chemists," Wiley-Interscience, New York, 1972, p. 81.
[11] S. S. Hirsh and M. R. Lilyquist, *J. Appl. Polym. Sci., 11,* 305 (1967).

SYNTHESIS AND CHARACTERIZATION OF RIGID-CHAIN POLYACENAPHTHYLENE

CHI-YU CHEN* and IRJA PIIRMA
*Institute of Polymer Science, The University of Akron,
Akron, Ohio 44325*

SYNOPSIS

Rigid-chain polyacenaphthylene (PAcN) was synthesized by a novel emulsion polymerization method that besides water and the sodium oleate emulsifier consisted of a water-soluble initiator, $K_2S_2O_8$, and the water-insoluble (at the polymerization temperature 50°), crystalline monomer acenaphthylene. The molecular weight of the emulsion PAcN was 150,000 by gel permeation chromatography (gpc). With the exception of a C^{13} nuclear magnetic resonance study, the PAcN was characterized in the bulk state by infrared spectroscopy and x-ray and electron diffraction. The C^{13} nmr spectrum in solution suggests a macromolecular association. The infrared spectra indicate that the PAcN can be either paracrystalline or amorphous. An amorphous infrared band at 670 cm^{-1} was identified. The emulsion PAcN is predominantly threo-disyndiotactic, and it is possible, therefore, to have a mesophase of nematic type in the bulk state. The paracrystallinity or ordered structure of PAcN is supported by wide-angle x-ray and electron diffraction patterns.

INTRODUCTION

Acenaphthylene (AcN) is a crystalline, aromatic vinyl monomer that has a bulky naphthalene moiety cis-disubstituted on the carbon-carbon double bond. The homopolymer of acenaphthylene was considered to be rigid by Kaufman and Williams [1] and Jones [2] on molecular structural grounds. There are four possible steroregular structures: the erythro-diisotactic, the erythro-disyndiotactic, the threo-diisotactic, and threo-disyndiotactic configurations. Story and Canty [3] have shown by model studies that only the threo-ditacticity structures can be constructed beyond trimers or tetramers. Dilute solution properties of high-molecular-weight polyacenaphthylene (PAcN) such as the second virial coefficient and the unperturbed mean-square end-to-end distance indicate that the molecule is unexpectedly flexible in solution [4, 5]. The flow birefringence of high-molecular-weight PAcN in bromoform solution demonstrates, however, that a skeletal rigidity does exist [6].

PAcN in solution has been shown to be degradable by light and heat, but less degradation has been observed in the solid state [7]. In this regard, it seems advantageous to study the properties and structure of PAcN in bulk. Of the bulk

* Present address: Department of Chemistry, The University of Alabama, University, AL 35486

Journal of Polymer Science: Polymer Symposium 65, 55–62 (1978)
© 1978 John Wiley & Sons, Inc.   0360-8905/78/0062-0055$01.00

properties so far only glass transition temperature and Young's modulus have been determined [8, 9]. The relatively high T_g and Young's modulus have been attributed to the high rigidity of this polymer in bulk [9].

For the synthesis of the rigid-chain PAcN, several methods have been reported [2-4, 10-12]. Cationic and anionic solution polymerizations usually yielded PAcN of lower molecular weight than is obtained by thermal free radical polymerization [3, 4]. Emulsion polymerization systems using organic solvents have also been reported [2].

This investigation included a novel emulsion polymerization of crystalline AcN and spectral studies of the corresponding polymer in bulk.

EXPERIMENTAL

Reagents

Commercial AcN (85% pure by nmr) was recrystallized from ligroin, sublimed in vacuo and then recrystallized from pentane. The final acenaphthylene was 98.6% pure with a melting point of 88°-90°. AcN in fine-powder form was obtained by coagulating the AcN-methanol solution in distilled water. The isolated AcN powder crystals were dried under vacuum at room temperature and used in the polymer synthesis.

Sodium oleate was prepared using purified oleic acid and methanolic sodium hydroxide solution. Potassium persulfate was used as received. The distilled water was additionally double-distilled.

Preparation of Polyacenaphthylene Latex

The ingredients were charged into a 240-ml narrow-mouthed bottle in the following order: 0.128 g $K_2S_2O_8$, 1.6 g sodium oleate, 8 g PAcN, and 72 g water. The metal bottle caps were lined with self-sealing butyl rubber gaskets. After purging with nitrogen, the polymerization bottles were clamped onto a rotating shaft in a thermostated water bath. They were rotated end-over-end at 40 rpm for 24 hr. The final latex was usually greenish-yellow, and conversions were about 85%. the PAcN latex was coagulated in methanol. The precipitated polymer was filtered through a Gooch crucible and dried in a vacuum oven at 50° for 2 hr.

Characterizations of Polyacenaphthylene

For the gpc and the spectroscopic characterization of the polymer the emulsion PAcN powder was dissolved in benzene and precipitated by the addition of methanol. The polymer was purified twice.

The molecular weight of the PAcN was determined by gel permeation chromatography, for which a Waters Associates Ana-Prep gel chromatograph was used. Tetrahydrofuran was used as the carrier solvent.

The infrared spectra of the polymer in KBr and KCl pellets and the polymer

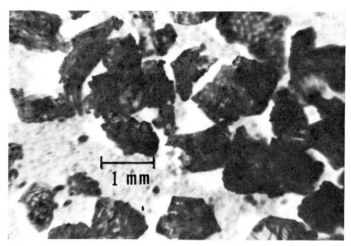

FIG. 1. Granular emulsion PAcN Powder.

film were obtained on a Perkin-Elmer 521 grating infrared spectrophotometer. The polymer film was cast on a NaCl plate from a 30% benzene solution.

The carbon-13 nmr spectra of 30% PAcN d-chloroform solutions were recorded at room temperature on a Varian Associates CFT-20 nmr spectrophotometer.

For the x-ray powder spectra, the polymer sample was pulverized first to fine powder and sieved through a screen. The wide-angle x-ray diffraction pattern was taken with a Philips Universal flat plate camera with a PW 1030 x-ray generator.

The electron diffraction pattern was obtained from a supported thin film of PAcN on a carbon grid (200 mesh). The polymer film was prepared by evapo-

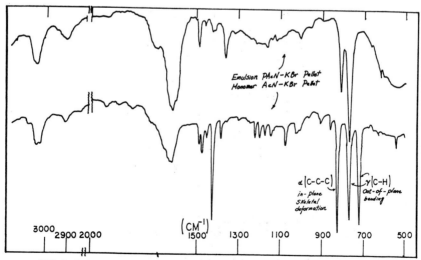

FIG. 2. Infrared spectra of emulsion PAcN and AcN Monomer in KBr pellets.

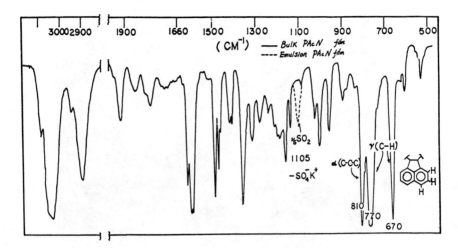

FIG. 3. Infrared spectra of cast PAcN films.

rating a 10% benzene solution of polymer on water surface. The mounted film
was examined with a JEW 120 transmission electron microscope.

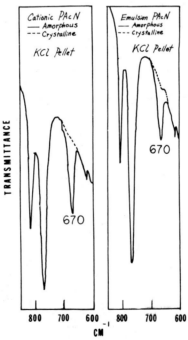

FIG. 4. Amorphous infrared band of PAcN at 670 cm⁻¹.

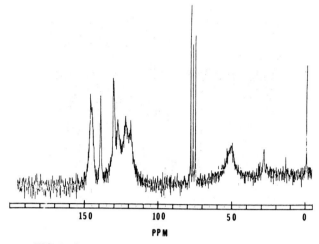

FIG. 5. C-13 nmr spectrum of emulsion PAcN in d-CHCl$_3$.

RESULTS AND DISCUSSION

Synthesis of Polyacenaphthylene

The unique feature of the emulsion polymerization of AcN as used in this work lies in the fact that the monomer is present as a crystalline solid during the polymerization; yet the rate of polymerization is not hampered or reduced. A preliminary discussion of this system has been reported by us [13], and a detailed kinetic and mechanistic evaluation will be published elsewhere [14].

The polymerization system was heterogeneous initially with micellar, aqueous, and monomer phases present. After high conversion was reached (above 65% conversion), the polymer latex became almost transparent and showed a greenish-yellow color. On coagulation in methanol, the latex separated into two

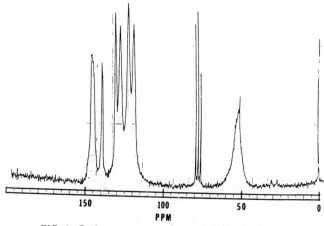

FIG. 6. C-13 nmr spectrum of cationic PAcN in d-CHCl$_3$.

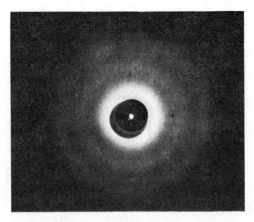

FIG. 7. Wide-angle x-ray pattern of emulsion PAcN powder.

distinct phases. The yellow solution contained unreacted monomer, emulsifier, and water; the other phase contained the precipitated polymer. The use of organic solvents to dissolve AcN monomer seems unnecessary as far as polymerization is concerned.

The dried PAcN thus obtained was granular and appeared microporous under the optical microscope. Figure 1 shows the irregular-shaped, sharp-edged granular polymer. This powder was further characterized, and the results are discussed in the following sections along with other comparable polymer samples.

The molecular weight of the emulsion PAcN was estimated from gel permeation chromatography and found to be 150,000 (weight average).

Infrared Spectra

Infrared analysis of PAcN in various forms revealed that the polymer can be either amorphous or crystalline. The polymers obtained from the precipitation of the latex or the polymer solution by adding a nonsolvent showed identical spectra (KBr pellet), as illustrated in Figure 2. The infrared spectrum of the monomer is also shown for comparison. The spectra of cast polymer films ex-

FIG. 8. Electron diffraction pattern of emulsion PAcN film.

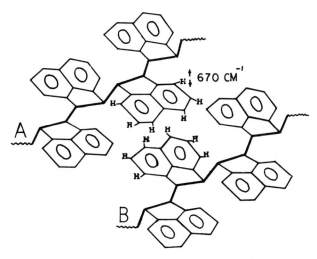

FIG. 9. Schematic ordered structure of PAcN.

hibited an additional band at 670 cm^{-1} band (Fig. 3) that we claim to be due to the amorphous nature of the polymer, as indicated by the transparency of these films. Additional evidence for this is provided by the fact that the polymer powder coagulated from latex or solution was shown to be paracrystalline by x-ray and electron diffraction patterns (discussed below) and did not exhibit the 670 cm^{-1} band. This amorphous band must be due to the out-of-plane bending mode of aromatic C—H. Apparently the bending mode is physically hindered by the ordered alignment of the macromolecules in their paracrystalline state.

The relationship between the paracrystalline state and the amorphous state is illustrated below:

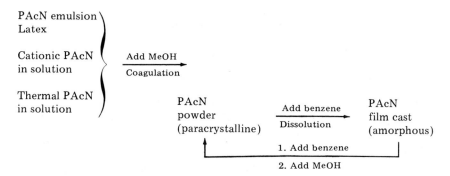

Therefore, it is believed that PAcN can assume predominantly either a paracrystalline or amorphous form, irrespective of the method of synthesis, i.e. whether by a free radical mechanism in emulsion or solution or by a cationic method. This is illustrated by infrared spectra in Figure 4.

According to Story and Canty [3], the 2895 cm^{-1} band is due to the threo-disyndiotacticity of this polymer, which leads us to conclude that the polymer prepared by emulsion polymerization has predominantly the same configuration.

Nuclear Magnetic Resonance Spectra

The C-13 nmr spectrum was obtained from emulsion PAcN in d-chloroform, as shown in Figure 5. Two features were especially noticed about the behavior of the emulsion PAcN sample in the magnetic field; namely, one was a low signal-to-noise ratio, as indicated by the spectrum, and the other was a greenish color shown by the polymer solution inside the magnetic probe. After checking another PAcN sample prepared by cationic method (diethylaluminum chloride initiator in hexane, molecular weight 25,000) that did not show the same color phenomenon (Fig. 6), we would have to speculate that a molecular association has occurred in the emulsion polymer sample. Similar association phenomena have been reported [15–17] for macromolecules such as poly(methyl methacrylate) and polyvinylchloride. Both the proton and carbon-13 nmr spectra are affected in such a way that some of the signals are usually "buried" in the noise. This is attributed to the immobility of the nuclei in the polymer lattice.

X-Ray and Electron Diffraction Patterns

The wide-angle x-ray and electron diffraction patterns of emulsion PAcN powder are shown in Figures 7 and 8, respectively. Qualitatively both patterns indicate a polymer of ordered structure, although high crystallinity is not indicated. Presumably a nematic mesophase in the bulk state could be visualized. The fact that ordered structure exists even though the polymer is amorphous by x-ray has been suggested recently by Yeh [18]. Since the emulsion PAcN is predominantly threo-disyndiotactic by infrared, the possible ordered structure is schematically shown in Figure 9.

REFERENCES

[1] M. Kaufman and A. F. Williams, *J. Appl. Chem., 1*, 489 (1951).

[2] J. I. Jones, *J. Appl. Chem., 1*, 568 (1951).

[3] V. M. Story and G. Canty, *J. Natl. Bur. Stand. A, Phys. Chem., 68A*, 2, 165 (1964).

[4] J. Moacanin, A. Rembaum, R. K. Laudenslager, and R. J. Alder, *J. Macromol. Sci. Chem., A1*, 1497 (1967).

[5] J. M. Barrales-Rienda and C. C. Pepper, *Polymer, 8*, 351 (1967).

[6] V. N. Tsvetkov, M. G. Vitovskaya, P. N. Lavrenko, E. N. Zakharova, I. F. Gavrilenko, and N. N. Stefanovskoya, *Polym. Sci. USSR, A13*, 2845 (1973).

[7] L. Utracki, N. Eliezer, and R. Simha, *J. Polym. Sci. Part B, 5*, 137 (1967).

[8] M. Morton, J. L. Trout, and T.-C. Cheng, *Proc. Int. Rubber Conf. Brighton, G6-1*, (1972).

[9] A. Yamata, M. Yanagita, and E. Kobayashi, *Rep. Inst. Phys. Chem. Res. Tokyo, 37*, 197 (1961).

[10] Von J. Springer, K. Ueberreiter, and R. Wenzel, *Makromol. Chem., 96*, 122 (1966).

[11] P. Lagroix and J.-C. Muller, *C. R. Acad. Sci. Ser. C, 264*, 1105 (1967).

[12] P. Giusti and F. Andruzzi, *Gazz. Chim. Ital., 96*, 1563 (1966).

[13] C. Y. Chen and I. Piirma, *Polym. Prepr. Am. Chem. Soc. Div. Polym. Chem., 16*(1), 234 (1975).

[14] I. Piirma and C. Y. Chen, in preparation.

[15] J. Spevacek and B. Schneider, *Makromol. Chem., 176*, 3409 (1975).

[16] Ch. E. Wilkes, *Macromolecules, 4*, 443 (1971).

[17] C. J. Carman, A. R. Tarpley, and J. H. Goldstein, *Macromolecules, 4*, 445 (1971).

[18] G. S. Y. Yeh, *J. Macromol. Sci. Phys., B*, 451 (1972).

THE POLYMERIZATION OF PHENYLACETYLENE IN THE PRESENCE OF FERRIC ACETYLACETONATE AND TRIETHYLALUMINUM

H. X. NGUYEN,* S. AMDUR,† P. EHRLICH, and R. D. ALLENDOERFER‡

Department of Chemical Engineering, State University of New York at Buffalo, Buffalo, New York 14214

SYNOPSIS

The polymerization of phenylacetylene initiated by ferric acetylacetonate and triethylaluminum was studied over the temperature range $-30°-70°$, in bulk as well as in the presence of diluents, and over a range of catalyst compositions and concentrations. The product consisted of four major fractions: crystalline polymer of molecular weight about 7000 and a very narrow molecular-weight distribution (MWD), (fraction I); somewhat lower-molecular-weight amorphous polymer with a broader MWD (fraction II); linear oligomer (fraction III); and the asymmetric cyclic triphenyl-benzene (fraction IV). The average degree of polymerization, of fraction I (65–70) is independent of all polymerization variables, that is, temperature from $-30°$ to $70°C$, the presence or absence of diluent (tetralin or benzene), and catalyst concentration and composition. The MWD is narrow with $\overline{M}_w/\overline{M}_n$ equal to 1.05 or less. The results suggest that thermodynamic rather than kinetic factors limit the size of the polymer. A possible mechanism is proposed.

INTRODUCTION

The polymerization of phenylacetylene in the presence of the catalyst system ferric acetylacetonate–triethylaluminum, $Fe(acac)_3-Al(C_2H_5)_3$, was described first by Noguchi and Kambara [1] and later by Kern [2]. Polyphenylacetylene (PPA-C) obtained in this manner shows crystallinity [3] and has a substantially higher molecular weight than PPA obtained by free radical, cationic, or thermal polymerization. We have reported some of the solution properties of this type of PPA [4].

The molecular conformation of the polymer appears to be like that of a rod or a stiff ribbon; polystyrene fractions with gpc elution volumes similar to those of the higher PPA fractions have molecular weights greater by an order of magnitude [4]. Conjugation and the phenyl substituents should make the polymer rather rigid. However, vibrational modes at microwave frequencies have been held responsible for facile downchain energy migration [5], and two di-

---

* Dow Chemical Co., Midland, Michigan.

† Wright State University, Dayton, Ohio.

‡ Department of Chemistry, State University of New York at Buffalo.

Journal of Polymer Science: Polymer Symposium 65, 63–71 (1978)
0360-8905/78/0062-0063$01.00

electric dispersion regions were observed in the *p*-methoxy and *p*-chloro derivatives of polyphenylacetylene. PPA-C is viewed best as a stiff molecule with a degree of rigidity as yet to be specified [6].

Like many conjugated polymers, PPA is paramagnetic in its pure form [7] and in solution [8, 9], and so the paramagnetism must be associated with molecular defects, which appear to occur quite generally in linear conjugated polymers. Such defects must be associated with bond alternation, which occurs in polymers of sufficient size according to most molecular orbital calculations [10], but no simple defect model appears to account for the paramagnetism [7, 9, 10]. Nonetheless, simple concepts of the electronic structure of growing polyene chains may account for peculiarities in the molecular weight and the molecular-weight distribution (MWD) of PPA.

Our previous studies showed that the MWD of the polymer was remarkable in its lack of a high-molecular-weight tail [4] when compared with condensation polymers or vinyl polymers obtained by other means than living anion processes. The rapid heterogeneous catalysis coupled with the insolubility of the polymer in the reaction mixture, which is characteristic of this polymerization, does not make it an easy subject for kinetic studies. An effort was therefore made to obtain information about the reaction mechanism by studying the effects of the polymerization variables on the molecular weight and the MWD of PPA polymerized in the presence of $Fe(acac)_3$–$Al(C_2H_5)_3$.

EXPERIMENTAL

The polymer was prepared as described in reference 4. On the basis of gross solubility characteristics of the polymer, we refer to parts A, B, and C, whereas components of the polymer, including oligomer, separable by further analytical techniques and concluded to correspond to different species according to degree of polymerization or structure are referred to as fractions I, II, III, and IV (see "Results"). Part A is insoluble in the reaction mixture and also in methanol; it contains mostly fractions I and II, a little IV, and a trace of III. Part B is polymer soluble in the hydrocarbon reaction mixture to which MeOH solubles, obtained by an exhaustive methanol wash of part A, have been added; it contains part of fraction II, most of fraction III, and a part of fraction IV. Part C is obtained by removal of part of the MeOH from the MeOH–hydrocarbon reaction mixture in which part B is insoluble, followed by acidification with HCl. Essentially pure fraction IV precipitates.

The gpc measurements were carried out as described [4], and the molecular weights reported are based on calibrations carried out with a Hitachi-Perkin Elmer Model 115 vapor pressure osmometer (vpo).

RESULTS

Figure 1 shows the gpc chromatogram of PPA-C made in the presence of tetralin at $-5°$, as obtained from the insoluble gel, which is the main reaction product (part A), after dissolution in *o*-dichlorobenzene at about 140°. Two

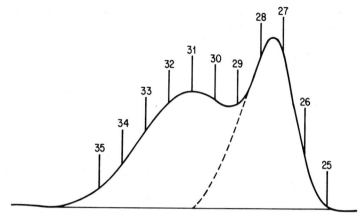

FIG. 1. Gpc chromatogram of polymer made in the presence of tetralin at −5°, part A.

major high-molecular-weight components are clearly indicated. If the polymerization is carried out at 30°, the lower-molecular-weight component appears as a broad shoulder [4]; at 70° it is far less pronounced (Fig. 2), as is also the case in the absence of diluent. Neither figure shows the contributions of the low-molecular-weight hydrocarbon soluble (part B) and the methanol soluble

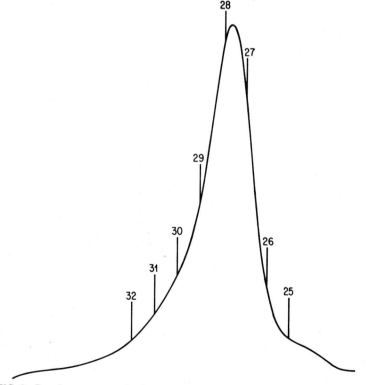

FIG. 2. Gpc chromatogram of polymer made in the presence of tetralin at 70°, part A.

components (part C). Recombination of parts A, B, and C for polymer prepared at 70° yields Figure 3, which shows the chromatogram resolved into four fractions. The molecular weights and distributions of these fractions were calculated according to the osmometric calibration already reported [4] and extended to lower molecular weights with PPA made in the presence of azo-bisisobutyronitrile [11]. The molecular weights of the four fractions are in decreasing order of M: fraction I with a mean molecular weight of 6500, fraction II with M mostly in the range of 2500–4000, fraction III with a mean molecular weight of about 800, and fraction IV, which was identified by nmr as the asymmetric cyclic trimer 1,2,4-triphenylbenzene. Fraction III has a molecular weight expected from PPA formed in the presence of conventional free radical initiators at the same monomer/diluent ratio and may be tentatively concluded to have been formed via the appropriate free radical mechanism [11]; it would then follow that fraction II is not formed in this manner. Fraction I has a molecular-weight distribution (MWD) that, when analyzed according to Figure 3, yields a value of $\overline{M}_w/\overline{M}_n$ of only 1.02, if no correction for boundary spreading is made. The MWD is narrower even than Figure 3 might suggest, because the elution volume is highly sensitive to molecular weight [4] and because boundary spreading probably makes an appreciable contribution to the chromatogram. We have already commented on the absence of a high-molecular-weight tail in PPA-C [4], but the resolution of high-molecular-weight PPA-C into fractions I and II renders the MWD of the former symmetrical and gives the corresponding chromatogram a near-Gaussian appearance.

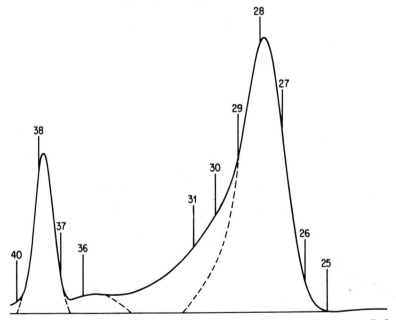

FIG. 3. Gpc chromatogram of polymer made in the presence of tetralin at 70°, parts A, B, C, recombined. Fractions I, II, III, IV between dotted lines.

Room-temperature soluble PPA prepared by Masuda et al. [12] in the presence of $WCl_6 \cdot \frac{1}{2}H_2O$ has a chromatogram virtually identical with that of fraction I (Fig. 4) and only a very slightly higher molecular weight. Higashimura (personal communication and ref. 12) reports an $\overline{M}_n$ of 12,400, whereas our measurements indicate 7200 (gpc, using the same calibration curve as that used for our samples) and 8200 (vpo); the differences are probably attributable to errors in osmometry by one or both parties. Whatever the correct absolute value of the molecular weight, we wish to emphasize the great similarity in chain length and MWD in linear PPA made with different transition metal catalysts.

Fraction I, as defined by the somewhat arbitrary graphical resolution shown, is closely similar to the crystalline PPA studied previously by x-ray diffraction [3], electron spin resonance [7], and electrical conductivity [13]. Recently Bloor obtained well-resolved resonance Raman spectra [14], after extracting part A, which contains mostly fractions I and II, with o-dichlorobenzene. This preparation, too, should be very similar to what we have defined as fraction I. As we have already shown, the MWD of fraction I is also very similar to soluble PPA obtained in the presence of WCl_6 [12]. It therefore seemed to be of particular interest to test the effect of reaction variables on the molecular weight of fraction I. Table I lists the reaction conditions and their variation in these experiments. The MWD is so narrow that a separate listing of $\overline{M}_n$, $\overline{M}_w$, and M corresponding to the peak count $\hat{M}$ is not very meaningful. Figure 5 shows the molecular weight of fraction I as the dependent variable. The temperature was chosen as the in-

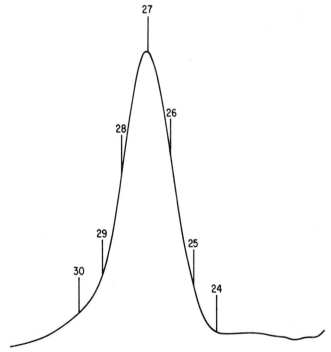

FIG. 4. Gpc chromatogram of PPA made in the presence of $WCl_6 \cdot \frac{1}{2}H_2O$. (Sample obtained through the courtesy of Professor Higashimura.)

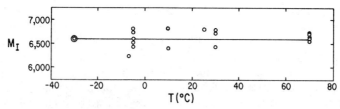

FIG. 5. Mean molecular weight M_I of fraction I as a function of polymerization temperature.

dependent variable, since one might have expected it, among all the reaction variables, to have the largest effect. Figure 5 shows that the polymerization temperature in the range $-30°-70°$ has no measurable effect on the molecular weight of the polymer. The points plotted in Figure 5 represent runs performed in the presence and absence of diluent (tetralin or benzene) and over a range of catalyst compositions and concentration. We note that none of the polymerization variables has an effect on the molecular weight of fraction I. Figure 6 shows the chain length distribution of fraction I, assuming corrections for axial dispersion need not be made.

TABLE I
Polymerization Conditions for Samples Whose M_I Is Plotted in Figure 5

Temperature (°C)	Diluent	x_M[a]	[Fe]/[M][b]	[Al]/[Fe][b]
−30	None	1.0	2.75×10^{-3}	2.85
−30	Tetralin	0.3	5.49×10^{-3}	2.5
−7	Tetralin	0.3	5.45×10^{-3}	2.5
−5	Tetralin	0.3	5.49×10^{-3}	2.5
−5	Tetralin	0.3	5.49×10^{-3}	2.5
−5	Tetralin	0.3	3.65×10^{-3}	2.5
−5	Tetralin	0.3	1.23×10^{-3}	7.5
−5	Tetralin	0.3	1.82×10^{-3}	5.0
10	Benzene	0.2	5.49×10^{-3}	2.5
10	Tetralin	0.3	5.49×10^{-3}	2.5
25	None	1.0	2.75×10^{-3}	2.85
30	Tetralin	0.3	3.65×10^{-3}	2.5
30	Tetralin	0.3	1.23×10^{-3}	7.5
30	Tetralin	0.3	1.82×10^{-3}	5.0
70	Tetralin	0.3	5.49×10^{-3}	2.5
70	Tetralin	0.3	1.82×10^{-3}	2.5
70	Tetralin	0.3	0.62×10^{-3}	7.5
70	Tetralin	0.3	5.49×10^{-3}	2.5
70	Tetralin	0.3	1.82×10^{-3}	5.0

[a] Initial mole fraction of monomer. A bulk run corresponds to an initial monomer concentration of 9.11 moles l^{-1}.

[b] Mole/mole.

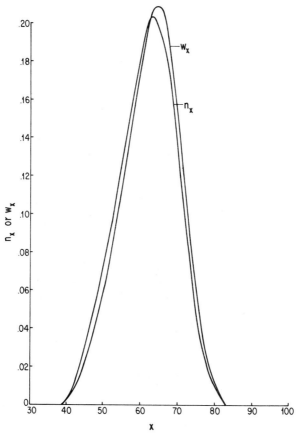

FIG. 6. Normalized number and weight chain length distribution functions n_x and w_x of fraction I. Uncorrected for axial dispersion.

DISCUSSION

The result that the molecular weight of fraction I is independent of all reaction variables and that the MWD can be characterized by such a small value of $\overline{M}_w/\overline{M}_n$ seems remarkable when viewed in terms of conventional mechanisms for vinyl and condensation polymerization. A living polymer mechanism can be ruled out because the nature of the catalyst should allow chain termination for which there is, in fact, evidence from the limited conversion and the progressive slowing down of the reaction. In particular, the insensitivity to temperature over the broad range $-30°C–70°C$ is difficult to rationalize as resulting from kinetic control of the reaction mechanism through competitive reactions with rate constants having an exponential temperature dependence, since the required close matching of activation energies is unlikely to occur over so broad a temperature range. At the same time, one must keep in mind the narrow MWD, which shows that either the propagation or the termination reaction must be strongly size-dependent. We would like to propose a possible mechanism to account for these observations.

According to molecular orbital treatments, which take into account σ-bond compression, bond alternation sets in at a critical chain length in cyclic even polyenes and odd linear polyene radicals [10, 15]. This argument can be extended to even linear polyene ion radicals. The origin of the greater stability of the asymmetric electronic structure—the one with bond alternation—over the symmetric one is a Jahn-Teller distortion and was shown by Peierls to be a very general property of one-dimensional solids with low-lying excited states [16]. One needs to postulate only that the growing chain has the characteristics of a delocalized π-radical, i.e., of a species without bond alternation, which changes to an asymmetric configuration with the unpaired electron at the chain end at a critical chain length, thus disrupting the transition metal-carbon bond. The critical chain length depends on the relative energies of two ground state electronic configurations and should be insensitive to temperature changes. Thus the distortion, with which we have associated the termination step, is temperature-independent. The model would then account for both the temperature-independent chain length and the sharp MWD. This type of explanation is, of course, quite different from that usually invoked for the termination of polyene chains of a shorter length, in which the cessation of chain growth is attributed to increasing opportunities for electron delocalization with increasing size, unaccompanied by an event that depends critically on size [17]. Although no strictly planar structure for PPA appears to be possible, several highly conjugated structures have been considered [7, 14, 18], and these include a cis-cisoid helix [2, 19].

Uncertainties pertaining to the degree of axial dispersion in the GPC measurements leave some doubt as to the true width of the MWD of fraction I, as does our recent observation that the chromatogram is broadened when a column with nominal pore size 10^5 Å is included in the column bank [20]. With columns 10^3, 10^4, 10^5, and 10^6 Å, $\overline{M}_w/\overline{M}_n$ of fraction I becomes 1.035, assuming a symmetrical MWD which is represented correctly by the high molecular weight half of the chromatogram and when no correction is made for axial disperson. Allowing for some error, we conclude that a value of 1.05 represents the upper limit of $\overline{M}_w/\overline{M}_n$ and that a correction for axial dispersion is likely to reduce that value to a smaller one [21].

We gratefully acknowledge support of this research by the National Science Foundation through its Polymers Program, Grant 7424522, and we greatly appreciate the gift by Professor Higashimura, Kyoto University, of PPA samples made in his laboratory.

REFERENCES

[1] H. Noguchi and S. Kambara, *Polym. Lett., 1,* 553 (1963).

[2] R. J. Kern, *J. Polym. Sci. Part A-1, 7,* 621 (1969).

[3] P. Ehrlich, R. J. Kern, E. D. Pierron, and T. Provder, *J. Polym. Sci. Part B, 5,* 911 (1967).

[4] B. Biyani, A. J. Campagna, D. Daruwalla, C. M. Srivastava, and P. Ehrlich, *J. Macromol. Sci. Chem., A-9,* 327 (1975).

[5] A. M. North, D. A. Ross, and M. F. Treadaway, *Eur. Polym. J., 10,* 411 (1974).

[6] A. M. North and P. J. Phillips, *Br. Polym. J., 1,* 76 (1969).

[7] G. M. Holob, P. Ehrlich, and R. D. Allendoerfer, *Macromolecules, 5,* 569 (1972).

[8] P. Ehrlich, E. C. Mertzlufft, and R. D. Allendoerfer, *J. Polym. Sci. Part B, 12,* 125 (1974).

[9] S. H. C. Chang, E. C. Mertzlufft, P. Ehrlich, and R. D. Allendoerfer, *Macromolecules, 8,* 642 (1975).

[10] L. Salem, "The Molecular Orbital Theory of Conjugated Systems," W. A. Benjamin, New York, 1966, Chap. 8.

[11] S. Amdur, A. T. Y. Cheng, J. C. Wong, P. Ehrlich, and R. D. Allendoerfer, *J. Polym. Sci. Part A-1, 16,* 407 (1978).

[12] J. Masuda, N. Sasaki, and T. Higashimura, *Macromolecules, 8,* 717 (1975).

[13] G. M. Holob and P. Ehrlich, *J. Polym. Sci. Part A-2, 15,* 627 (1977).

[14] D. Bloor, *Chem. Phys. Lett., 43,* 270 (1976).

[15] H. C. Longuet-Higgins and L. Salem, *Proc. R. Soc., A-251,* 172 (1959).

[16] R. E. Peierls, "Quantum Theory of Solids, Oxford, 1955, p. 108.

[17] J. Manassen and R. Rein, *J. Polym. Sci. Part A-1, 8,* 1403 (1970); J. Kriz, *ibid. Part A-2, 10,* 615 (1971).

[18] D. Bloor and O. Rohde, *Chem. Phys. Lett.,* (1978), to appear.

[19] C. I. Simionescu, V. Percec, and S. Dumitrescu, *J. Polym. Sci., Polym. Chem. Ed., 15,* 2497 (1977).

[20] We are indebted to R. C. Gebauer and E. Slagowski, Hooker Chemical Co., for this measurement.

[21] H. S. Chi and P. Ehrlich, unpublished results.

MOLECULAR MECHANISMS IN THE FORMATION OF POLYDIACETYLENES IN THE SOLID STATE

J. B. LANDO and D. DAY
Department of Macromolecular Science, Case Western Reserve University, Cleveland, Ohio 44106

V. ENKELMANN
Institut fur Makromolekulare Chemie, der Universitat Freiburg, D7800 Freiburg, Germany

SYNOPSIS

Crystals of poly(5,7-dodecadiinediol-1,12-bisphenylurethane) are monoclinic, $P2/_1a$. The crystal structure has been determined and refined by full matrix least-squares analysis of diffractometrically measured intensities. The final R index is 0.081 (719 observed reflections). Intramolecular hydrogen bonding, a strong contribution of electron delocalization in the polymer backbone, and unusual C—C bond lengths in the side chain are observed. Polymerized crystals of o,o'-bisphenyl glutarate diacetylene crystallize in a C2/c monoclinic space group. The x-ray refinement reveals a solid solution of the monomer and polymer. It was determined that the only movement that occurs during polymerization is in the diacetylene rod (a rotation of 33°) and in adjacent phenyl groups (a rotation of 19° and a swing of 6°) and all other atomic positions in the side group remain essentially the same. An infrared investigation and the final refined atomic coordinates with their standard deviations indicate the BPG crystal was polymerized to an extent of 35%. The final R index is 0.075 (902 observed reflections). Again a strong contribution of electron delocalization in the polymer backbone is observed.

INTRODUCTION

Many substituted diacetylenes have been found to be highly reactive in the solid state [1]. Topochemical polymerization is initiated by exposure to ultraviolet, x-ray, or gamma radiation or by annealing. Each monomer molecule joins with two neighboring molecules in a 1,4-addition reaction at the conjugated triple bonds to form a linear fully conjugated polymer chain. In this way large, nearly defect-free polymer single crystals can be obtained [1]. During such a reaction unit-cell dimensions show only slight changes, and the space group is retained. Thus a solid solution exists at all conversions [1]. From x-ray diffraction [2, 3] (R=H_2C—C—CO—NH—ϕ, R=H_2C—O—SO_2—ϕ—CH_3) and Raman spectral studies [4] the polymer backbone is best represented by the mesomeric structure I (acetylenic) in Figure 1. However, there is in some

Journal of Polymer Science: Polymer Symposium 65, 73–78 (1978)
© 1978 John Wiley & Sons, Inc.   0360-8905/78/0065-0073$01.00

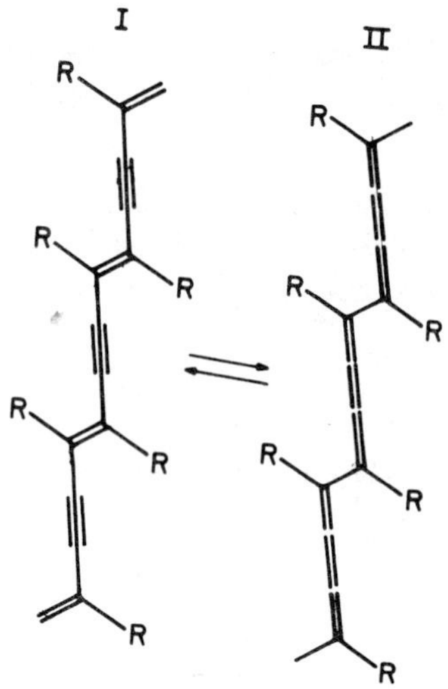

FIG. 1. Acetylenic (I) and butatriene (II) chain structures.

polymers spectroscopic evidence of significant resonance contribution from structure II (butatriene) in Figure 1, corresponding to considerable π-electron delocalization [4].

The structures we investigated were poly(5,7-dodecadiinediol-1,12-bis-phenylurethane) poly(TCDU) ($R=(-CH_2)_4-O-CONH-\phi$) and a solid solution at maximum conversion of o,o'-bisphenyl glutarate diacetylene (BPG) and poly(o,o'-bisphenyl glutarate diacetylene) poly(BPG) ($2R=\phi-O-CO-(CH_2)_3-OC-O-\phi$). Primary interactions between side chains in these polymers—hydrogen bonding in poly(TCDU), covalent bonding in the cyclic poly(BPG)—should be intramolecular in contrast to the strong intermolecular interactions observed in earlier polydiacetylene crystal structures [2, 3]. It was expected that this difference might affect the electronic structure of the polymer backbones. In addition the existence of a stable solid solution of (BPG) and poly(BPG) indicated the possibility that a complete crystal structure would yield detailed information concerning molecular motions necessary for polymerization.

EXPERIMENTAL

Poly(TCDU) crystals were prepared from crystals of the corresponding monomer by gamma-ray irradiation with a dose of 100 Mrads. The specimen used for intensity measurement was $0.6 \times 0.3 \times 0.1$ mm and was mounted along a. The systematic extinctions were 0k0 with k odd and h0l with h odd. Thus the space group is $P2_1/a$. The unit cell dimensions are $a = 6.229$ Å, $b = 39.03$ Å, $c = 4.900$ Å (chain axis), $\beta = 106.85$, $D_{calc} = 1.257$ g/cc, and $Z = 4$.

Single crystals containing solid solutions of poly(BPG) and BPG were prepared from BPG crystals by gamma-ray irradiation with a dose of 60 Mrads.

The specimen used for intensity measurement was $1.0 \times 0.4 \times 0.1$ mm and was mounted along c. Systematic absenses were hkl with h + k odd and h0l with l odd, indicating a space group of C2/c or Cc. The final determined structure gave a C2/c space group. Unit cell dimensions are a = 23.12 Å, b = 7.87 Å, c = 9.69 Å (chain axis), β = 111.29°, D_{calc} = 1.34 g/cc, and Z = 4.

Intensity data were collected using a Picker Facs-I system, consisting of a four-circle diffractometer controlled by a PDP-8/1 computer, Ni-filtered copper Kα x-ray radiation was used. Intensities of 1140 (TCDU) 2600 (BPG) independent reflections within the limiting sphere of 2θ = 110° were measured using the 2θ scan mode, scanning from 1.5° (TCDU) or 1.0° (BPG) below to 1.5° (TCDU) or 1.0° (BPG) above the calculated Bragg angle at a rate of 1° (TCDU) or 2° (BPG) per minute. Background counts were made for 20 (TCDU) 10 (BPG) sec. Three reflections ($\overline{1}$50, 060, 001) (TCDU) (10,00,020 and 008) (BPG) were used as standards and were scanned at 50 reflection intervals. The standards showed no significant change during the data collection. The raw data were corrected for Lorentz and polarization effects and absorption. 724 (TCDU) 902 (BPG) reflections with intensities larger than two times (TCDU) 2.5 times (BPG) their standard deviation were considered to be observed. The others were given zero weight during the refinement.

STRUCTURE DETERMINATION

The crystal structures were determined by direct methods. All nonhydrogen atoms could be readily located in the E-maps. Refinement was by full matrix least-squares calculations. The hydrogen atoms were assigned physically reasonable coordinates, and the temperature factors of the corresponding nonhydrogen atoms and their contributions were included in structure factor calculations. On termination of the refinement R (for observed reflections only) for poly(TCDU) was 0.081 and 0.075 for the solid solution of poly(BPG) and BPG.

The bc projection of poly(TCDU) is shown in Figure 2. Although not indicated in this figure, reasonable anisotropic temperature factors were obtained for all atoms. Hydrogen atom temperature factors were held isotropic. Standard deviations for bond angles were 1°, and standard deviations for bond lengths were 0.01–0.02 Å. The polymer backbone is planar within these limits.

The ac projection of one polymer chain and the corresponding BPG monomer stack are shown in Figure 3. Thermal ellipsoids and hydrogen atoms are indicated in this figure. Only four atoms in the asymmetric unit of structure (half a monomer unit) have sufficiently different positions (over 0.4 Å) to be assigned a separate position in the monomer and polymer structures. These were the two atoms of the diacetylene rod and two of the phenyl group carbons (the carbon bonded to the rod and the carbon para to it). Of these four atoms, only the atom between bonds B and C were sufficiently far apart to refine anisotropic temperature factors. The other three carbon atoms along with all hydrogen atoms were held isotropic. The large anisotropic thermal ellipsoids of three of the remaining four carbon atoms in the phenyl group reflect their unresolvable

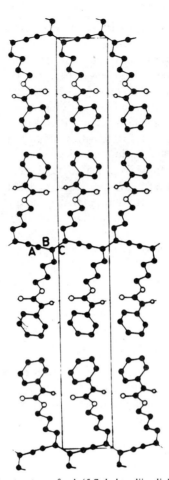

FIG. 2. bc projection of the structure of poly(5,7-dodecadiinediol-1,12-bisphenylurethane).

movement on polymerization. The remainder of the molecule had almost
identical positions in the monomer and polymer indicated by normal thermal
ellipsoids. Standard deviations of bond angles in this structure were around 1°,
and standard deviations of bond lengths for the monomer were 0.01–0.02 Å and
for the polymer 0.01–0.03 Å. On the basis of infrared spectra on the one hand
and the minimization of the residual and standard deviations on the other, the
solid solution was found to contain 35% polymer and 65% monomer. Ring strain
causes the diacetylene rod of the monomer to bend 3° and a nonplanarity of the
connecting bond of the polymer backbone of 3°.

DISCUSSION

In the two structures we have completed, both having strong intramolecular
interactions between side chains, the bond lengths of the backbone bonds, labeled
A, B, and C in eq. (1) and Figures 2 and 3(b), are for poly(TCDU) A = 1.24

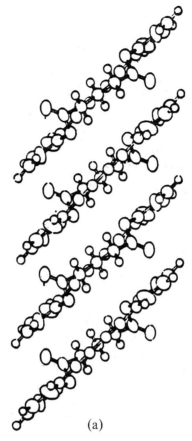

(a)

FIG. 3. ac projection of the structure of (a) o,o'-bisphenyl glutarate diacetylene.

Å, B = 1.37 Å, C = 1.44 Å (a second refinement gave A = 1.17 Å, B = 1.38 Å, C = 1.46 Å) and for poly(BPG) A = 1.29 Å, B = 1.38 Å, C = 1.42 Å. The values indicate some electron delocalization but a stronger contribution of the butatriene structure—form II in eq. (1)—in sharp contrast to the acetylenic structure found for polydiacetylenes having primarily intermolecular side-chain interactions A = 1.21 Å, B = 1.41 Å, C = 1.36 Å [2] and A = 1.19 Å, B = 1.43 Å, C = 1.36 Å [3].

From the structure of the solid solution of BPG and poly(BPG), if we make the reasonable assumption that the structures of the pure materials are changed little in the solid solution, a clear picture of the molecular motions necessary for reaction is obtained. The diacetylene rod rotates 33°, and the phenyl group rotates 19° and turns in its plane 6°, allowing the rest of the molecule to maintain the same position.

Finally a peculiar feature of the poly(TCDU) structure must be mentioned. The carbon-carbon bonds of the side chain have a sequence of bond lengths of 1.56, 1.43, 1.58, 1.42 Å with slightly enlarged bond angles. Since a second refinement of the data gave virtually identical results and the thermal ellipsoids

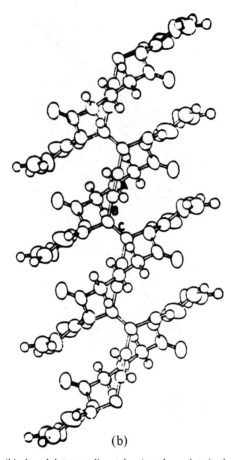

(b)

Fig. 3. (b) Poly(bisphenylglutarate diacetylene) as they relate in the solid solution.

appear reasonable, this effect is presumably real. What this apparent induced conjugation arises from is not clear.

We are grateful for the partial support of this work by NATO research grant No. 913 and by a NATO postdoctoral research grant to V. Enkelmann. Helpful discussions with J. Scheinbeim, R. H. Baughman, and G. Wegner are also gratefully acknowledged.

REFERENCES

[1] G. Wegner, *Adv. Chem. Ser., 129,* 255 (1973).
[2] E. Hadiche, E. C. Mez, C. H. Krauch, G. Wegner, and J. Kaiser, *Angew. Chem., 83,* 253 (1971).
[3] D. Kobelt and H. Paulus, *Acta Cryst., B30,* 232 (1974).
[4] A. J. Melveger and R. H. Baughman, *J. Polym. Sci. Part A-2, 11,* 603 (1973).

DEPOLARIZED RAYLEIGH SPECTROSCOPY OF RIGID MACROMOLECULES

R. PECORA

Department of Chemistry, Stanford University, Stanford, California 94305

SYNOPSIS

Some applications of depolarized light-scattering spectroscopy to the study of rigid macromolecules in solution are described. The depolarized light-scattering method can be used to obtain rotational diffusion coefficients and static and dynamic correlation factors of rigid macromolecules. In molecules exhibiting some flexibility, intramolecular conformational relaxation times may also be obtained. Dynamic polarized light-scattering measurements give the orientation averaged translational diffusion coefficient. With the Perrin relations for spheroids, the rotational and translational diffusion coefficients may be combined to yield the macromolecular solution dimensions. In addition, changes in rigidity and shape of rigid macromolecules may be observed by measuring changes in the depolarized spectrum. For instance, the total depolarized intensity and relaxation time usually decrease when a helical molecule is denatured to a flexible form. The rotational diffusion coefficients obtained by light scattering may also be combined with relaxation times obtained by other techniques such as nuclear magnetic relaxation to draw inferences about molecular rigidity and local motions.

INTRODUCTION

Depolarized light-scattering spectroscopy has become a major tool for investigating the orientational correlations and rotational dynamics of small molecules in the liquid state [1]. Systematic studies of dynamic and static orientation factors as well as the viscosity dependence of "single-molecule" reorientation times have been performed. The latter studies have, for instance, led to the introduction of hydrodynamic models using "slip" (or "no-stick") boundary conditions to describe reorientational motion of single molecules in liquids [2, 3]. By contrast, however, little work has been done on depolarized light-scattering spectroscopy of macromolecules in solution [4–12]. Much of the dynamic light-scattering work on macromolecules has used the polarized scattering to obtain macromolecular translational diffusion coefficients [13]. Although the depolarized signals are usually much weaker than the polarized ones, it is clear from previous work that their strength is usually adequate for proper measurement. In this article, we describe some of the applications of depolarized light-scattering spectroscopy to the study of rigid macromolecules in solution.

Journal of Polymer Science: Polymer Symposium 65, 79–84 (1978)
0360-8905/78/0065-0079$01.00

THEORETICAL

Consider a solution of rigid, optically anisotropic molecules. The molecules are assumed to be small compared with the reciprocal of the scattering vector length q^{-1} so that intramolecular interference factors may be ignored in the depolarized scattering. A rough criterion for this neglect is that $qL < 5$, where L is a characteristic dimension of the macromolecule. This condition may in many cases be experimentally obtained even for fairly large molecules by observing the depolarized scattering at low scattering angles, since q depends on the scattering angle θ by the relation $q = (4\pi/\lambda) \sin (\theta/2)$, where λ is the wavelength of the incident light in the scattering medium. For simplicity it is furthermore assumed that the molecules are cylindrically symmetric with polarizability anisotropy β. The frequency spectrum of depolarized light scattered from this system may be shown to be [14, 15]

$$I_{VH}(q,\omega) = \frac{AN}{15\pi}\beta^2(1 + Nf)\frac{(1/\tau_{LS})}{(\omega^2 + 1/\tau_{LS}{}^2)} \tag{1}$$

where A depends on the optical constants of the medium and the incident intensity, N is the number of macromolecules per unit volume, f is a measure of the average static orientational correlation between two scatterers, and τ_{LS} is the light-scattering relaxation time.

The light-scattering relaxation time is given by

$$\frac{1}{\tau_{LS}} = q^2D + 6\Theta \left(\frac{1 + Ng}{1 + Nf}\right) \tag{2}$$

In eq. (2), D is the macromolecular translational diffusion coefficient in a volume-fixed frame of reference. The quantity Θ is the rotational diffusion coefficient of the symmetry axis of a single macromolecule, and the factors f and g correct, respectively, for static and dynamic orientational correlations. The static correlation factor f also appears in the equation for the frequency-integrated scattering

$$I_{VH} = \frac{AN\beta^2}{15} (1 + Nf) \tag{3}$$

and so may be independently obtained from integrated intensity measurements. For many macromolecular systems the translational contribution to $1/\tau_{LS}$ is negligible, or may be made negligible by observing the spectrum at low q, and $1/\tau_{LS}$ is determined entirely by the second term on the right-hand side of eq. (2). Furthermore, in the limit of infinite dilution, the terms containing f and g are negligible and $1/\tau_{LS} = 6\Theta$. Since the viscosity dependence of Θ is usually known, the value of Θ can then be extrapolated to more concentrated solution conditions, and this extrapolated value of Θ and the f obtained from integrated intensity measurements can be combined in eq. (2) to yield values of g. Thus, the rotational diffusion coefficient Θ and the static and dynamic orientation factors f and g may be directly obtained. Values of f and g are especially important in studying the onset of the anisotropic phases of solutions of rigid polymers.

It is well known that the rotational diffusion coefficient of macromolecules in solution may be calculated from hydrodynamics using "stick" boundary conditions. For ellipsoids of revolution, the familiar Perrin equations may be used to express the rotational diffusion coefficient of the major axis in terms of the solution viscosities and the lengths of the major and minor axes [16].

It is now a relatively routine matter to measure translational diffusion coefficients of macromolecules in solution by polarized light scattering using autocorrelation or equivalent techniques. Since the average translational diffusion coefficient of an ellipsoid of revolution may be related to the ellipsoid dimensions [16], depolarized and polarized light scattering may be combined to yield the macromolecule solution dimensions [5].

It is also relatively simple to study conformational changes from rigid to flexible states. For instance, the depolarized integrated intensity and relaxation time both decrease as a helical molecule is denatured to a coil form at constant solvent viscosity.

The results outlined above for rigid, cylindrically symmetric molecules may be generalized to rigid macromolecules of arbitrary shape. In the most general case, the depolarized spectrum consists of a sum of five Lorentzians instead of the simple single-Lorentzian form of eq. (1) [17]. In the event that the principal axes for the polarizability and rotational diffusion tensors coincide, the depolarized spectrum reduces to the sum of only two Lorentzians. Analyzing these multiple-Lorentzian forms, however, is very difficult in practice, especially in macromolecular solutions that exhibit polydispersity. Successful analysis of these systems requires that the time constants for the Lorentzians be very different. A combination of light scattering and other techniques that measure molecular reorientation times such as carbon-13 nuclear magnetic relaxation in many cases permits analysis of the anisotropic rotation of asymmetric molecules [15].

Some Applications

Depolarized scattering experiments are performed using apparatus described in detail elsehwere [11]. For "fast" processes (relaxation times less than about 10^{-6} sec), Fabry-Perot interferometry is used to measure the spectrum of scattered light. For slower processes, autocorrelation of the scattered light intensity is used to obtain the relaxation times directly. Autocorrelation is also used to obtain translational diffusion times from polarized light-scattering intensities.

The three experiments briefly described below illustrate some of the uses of depolarized light-scattering spectroscopy mentioned under "Theoretical."

Intramolecular Relaxation Times of Flexible Polymers [8, 9, 11]

Depolarized spectra of flexible and semiflexible polymers yield intramolecular relaxation times. For example, the depolarized spectra of polystyrenes in dilute solution exhibit two components—a slow component whose relaxation time is proportional to the molecular weight and a fast component whose relaxation time is independent of molecular weight [11]. For dilute polystyrenes in CCl_4, the fast relaxation time is 4.5 ± 1.0 nsec. The slow relaxation time in these systems is identified with half of the relaxation time of the longest Rouse-Zimm mode for the chain, and the fast, molecular-weight-independent time is assigned to a local, correlated motion of phenyl groups about the main chain backbone. Thus, dynamic constants for both "local" and "long-range" intramolecular motions may be obtained for polymers of this type.

Determination of Rigidity of a Protein Backbone [10]

Depolarized light scattering gives the overall molecular rotational diffusion coefficient for rigid macromolecules. Carbon-13 nuclear magnetic longitudinal relaxation times under certain circumstances yield the reorientation time of the carbon–proton bond in a molecule. This carbon–proton bond reorientation time depends both on overall molecular reorientation and local bond motion. It is unlikely that local carbon–hydrogen bond motion would affect the depolarized light-scattering spectrum. Thus, agreement of the light-scattering relaxation time and that for reorientation of a particular carbon–hydrogen bond is strong evidence that the carbon–hydrogen bond relaxation is due to overall molecular reorientation and contains negligible contributions from local motions.

Both depolarized light scattering and carbon-13 nuclear magnetic relaxation experiments were performed on muscle calcium-binding protein (MCBP) isolated from carp. MCBP is a protein of molecular weight 12,000. X-Ray studies of the crystal show it to be a prolate ellipsoid with approximate dimensions 30 $\times$ 36 Å. The α-carbons of the amino acids in this protein form a homogeneous group whose reorientation time was found by NMR to be in the range of 11–14 nsec. The depolarized light-scattering spectrum gave a single Lorentzian whose width gave a molecular reorientation time of 12 ± 1 nsec. This similarity implies that the α-carbons in MCBP form a rigid backbone that rotates with the whole molecule.

Solution Dimensions of Macromolecules [5, 18]

As mentioned under "Theoretical," depolarized light scattering can be used to measure the rotational relaxation time of a molecule, and polarized scattering can measure the translational diffusion coefficient. Such measurements have been performed on the bacterial antibiotic gramicidin A [18]. This molecule is a linear pentadecapeptide with alternating L and D amino acid residues. It induces specific transport of alkali cations through biological and model membrane systems. In addition, it is thought to form a dimer of length about 40–50 Å in membranes [19]. The conformation of gramicidin A in various solvents has been studied by a variety of techniques, including carbon-13 nmr [20].

Depolarized and polarized light-scattering spectra of gramicidin in methanol, ethanol, and dimethylsulfoxide (DMSO) have been obtained in our laboratory [18]. Measurements of the rotational relaxation in ethanol (3.9 nsec) indicate that the value previously determined by Fossel et al. by carbon-13 nmr is in error

[20]. By combining the measured translational and rotational diffusion coefficients and using the Perrin equations for ellipsoids, the solution dimensions of the molecule may be obtained. In methanol, we determined that (solvated) gramicidin is an ellipsoid with major axis 41 Å and minor axis 20 Å. In ethanol these numbers are 47 Å and 15 Å, respectively. Both sets of numbers indicate the presence of dimers in the solutions. The size discrepancy probably arises from the presence of different amounts of higher aggregates in the two solvents. Studies of the concentration dependence of the depolarized spectra should elucidate this point. In DMSO, gramicidin A is thought to be a flexible coil [20]. The depolarized light-scattering signal is, in this case, very weak, as might be expected. It appears that in DMSO we are observing intramolecular motions of the flexible gramicidin in the depolarized spectrum. This experiment, however, is being redone at higher gramicidin concentrations in order to obtain better signal-to-noise ratios.

DISCUSSION

Some illustrations of the use of depolarized light scattering in the study of macromolecules have been given. The total number of experiments that have been performed to date in this field is very small, possibly because few polymer scientists performing light scattering are as yet equipped with Fabry-Perot interferometers. Most dynamic light-scattering experiments on polymers in solution have used autocorrelation techniques exclusively. This reliance on autocorrelation techniques restricts the applicability of depolarized spectroscopy to the study of slow relaxation times of very high-molecular-weight polymers. Both autocorrelation and interferometry should be used to obtain maximum applicability of the light-scattering technique.

Many of the potential applications of depolarized light scattering to polymers in solution have not as yet even been attempted. For instance, no studies of the static and dynamic correlation factors for macromolecules mentioned under "Theoretical" have been published. Another unexplored possibility is the use of resonance enhanced light depolarized scattering to study molecules in very dilute solution or to study local motions of macromolecules [21].

REFERENCES

[1] D. R. Bauer, J. I. Brauman, and R. Pecora, *Ann. Rev. Phys. Chem.*, 27, 443 (1976).
[2] C. Hu and R. Zwanzig, *J. Chem. Phys.*, 60, 4354 (1974).
[3] D. R. Bauer, J. I. Brauman, and R. Pecora, *J. Am. Chem. Soc.*, 96, 6840 (1974).
[4] A. Wada, N. Suda, T. Tsuda, and K. Soda, *J. Chem. Phys.*, 50, 31 (1969).
[5] S. B. Dubin, N. A. Clark, and G. B. Benedek, *J. Chem. Phys.*, 54, 5158 (1971).
[6] J. M. Schurr and K. S. Schmitz, *Biopolymers, 12,* 1021 (1973).
[7] T. A. King, A. Knox, and J. D. G. McAdam, *Biopolymers, 12,* 1917 (1973).
[8] K. S. Schmitz and J. M. Schurr, *Biopolymers, 12,* 1543 (1973).
[9] C.-C. Han and H. Yu, *J. Chem. Phys., 61,* 2650 (1974).
[10] D. R. Bauer, S. J. Opella, D. J. Nelson, and R. Pecora, *J. Am. Chem. Soc., 97,* 2580 (1975).
[11] D. R. Bauer, J. I. Brauman, and R. Pecora, *Macromolecules, 8,* 433 (1975).
[12] D. R. Jones and C. H. Wang, *J. Chem. Phys., 65,* 1835 (1976).

[13] See, for instance, B. J. Berne and R. Pecora, "Dynamic Light Scattering with Applications to Chemistry, Biology and Physics," Wiley, New York, 1976.

[14] T. Keyes and D. Kivelson, *J. Chem. Phys., 56,* 1057 (1972).

[15] G. R. Alms, D. R. Bauer, J. I. Brauman, and R. Pecora, *J. Chem. Phys., 59,* 5310 (1973); *61,* 2255 (1974).

[16] F. Perrin, *J. Phys. Radium, 5,* 497 (1934); *7,* 1 (1936).

[17] R. Pecora, *J. Chem. Phys., 49,* 1036 (1968).

[18] Q.-H. Lao and R. Pecora, in preparation.

[19] M. C. Goodall, Biochim. *Biophys. Acta, 219,* 28, 470 (1970).

[20] E. T. Fossel, W. R. Veatch, Y. A. Ouchinnikov, and E. R. Blout, *Biochemistry, 13,* 5264 (1974).

[21] D. R. Bauer, B. Hudson, and R. Pecora, *J. Chem. Phys., 63,* 588 (1975).

NEMATIC POLYMERS

P. PINCUS*

Department of Physics, University of California, Los Angeles, California 90024

P. G. DE GENNES

College de France, 75231 Paris, France

SYNOPSIS

We present a primitive model that describes a cooperative helix-coil liquid–crystal phase transition. Consider a solution of macromolecules that individually undergo a helix-coil "transition" as a function of some external parameter such as temperature or pH. At some given value of the parameter, the chain is composed of a statistical distribution of rigid (helical) segments and flexible coiled segments. We then suppose that there exists an intermolecular Maier-Saupe type of quadrupole-quadrupole interaction between the rigid sections that derives from excluded volume and van der Waals forces. In terms of a mean-field approximation, we investigate the possible phase transitions as a function of polymer concentration. In particular, we show that the intermolecular coupling may lead to a first-order transition in which the macromolecules make a transition from a nearly coiled to a nearly rigid conformation accompanied by the simultaneous development of a long-range nematic type of liquid-crystalline orientational order.

INTRODUCTION

There exist some preliminary indications [1] that certain solutions of flexible linear macromolecules may undergo a phase transition, e.g., as a function of monomer concentration, in which the individual molecules become rigid simultaneously with the development of mesmorphic order. The purpose of this study is to present a simple model that describes a situation in which intermolecular steric interactions between rigid segments generate a nematic phase of nearly rigid macromolecules. Under "The Model," we introduce the model, which we develop under "Mean-Field Theory." Under "Discussion," we discuss our results and present some shortcomings that should be investigated in future studies.

* Supported in part by the National Science Foundation and by the Office of Naval Research.

THE MODEL

We consider an extremely dilute solution of macromolecules that individually undergo a helix-coil transition as a function of external parameters such as temperature or pH. At fixed values of these external quantities the conformational state of the macromolecule is determined by two parameters: (1) a Boltzmann factor $s = \exp(-\Delta F/k_B T)$, where ΔF is the difference in free energy per monomer between units in randomly coiled and in rigid configurations; (2) $\sigma^{1/2} = \exp(-F_w/k_B T)$, where F_w is the free energy necessary to create a boundary between rigid and coiled segments. The origin [2] of the wall energy is usually associated with the finite range along the chemical sequence of the hydrogen bonding interaction. The width of the transition region between the helical and coiled limits as a function of s becomes smaller as σ becomes smaller; i.e., as the energy required to form a boundary increases, the transition becomes more cooperative.

If the external parameters that affect the helix-coil transformation are fixed, the molecular conformation may be pictured as a chain of helical domains separated by coiled segments. The key assumption of our model is to suppose that there exists a Maier-Saupe [3] interaction between the rigid (helical) segments on different molecules. This interaction, which has the symmetry of the coupling between quadrupoles, presumably has its origin in a combination of steric repulsion and van der Waals attraction. It is this type of interaction which leads to the nematic liquid–crystalline structure. In the mean-field approximation, the Maier-Saupe (M-S) coupling leads to an additional force favoring the helical domains, which is proportional to the overall monomer concentration c. If the interaction is sufficiently strong, a phase transition may occur, e.g., as a function of c, in which the coupling drives a real first-order helix-coil transition simultaneously with the development of long-range orientational correlations. In the next section, we develop the molecular field theory.

MEAN-FIELD THEORY

We first review a simplified version [2] of the standard helix-coil transition theory for an isolated macromolecule. Assuming a polymerization index $N \gg 1$, such that end effects are negligible, we write the partition function as $Z_N = \lambda^N$, where $(-k_B T \ln\lambda)$ is the free energy per monomer. In order to compute λ, let us imagine adding one coil region of n monomers and one helical region of m monomers to the existing chain of N-$(m + n)$ units (Fig. 1).

$$\lambda^N = \sum_{m,n=1}^{N} \lambda^{N-(m+n)} \sigma s^m \tag{1}$$

where the factor σ is associated with adding two boundaries and there is a factor s for each hydrogen-bonded monomer. This is a reasonable approximation for long chains if we are not too close to the limiting states of one large coil or one

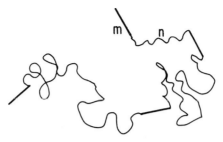

FIG. 1. A macromolecule during a helix-coil transition exhibiting rigid fluctuations.

rigid helix. Performing the sums, we obtain the eigenvalue equation for λ:

$$(\lambda - s)(\lambda - 1) = \sigma s \qquad (2)$$

where the physical root corresponds to the largest value of λ. The two extreme limits are (1) no intrinsic cooperativity where $\sigma = 1$, i.e., no additional wall energy, $\lambda = 1 + s$; (2) cooperative limit where $\sigma \to 0$, i.e., the wall energy is dominant leading to $\lambda = 1$ or $\lambda = s$, whichever is larger. Clearly $\lambda = 1$ represents a completely coiled state and $\lambda = s$, a completely rigid conformation. A convenient order parameter is the fraction Θ of monomers in the rigid segments where

$$\Theta = \left(\frac{s}{\lambda}\right)\left(\frac{\partial \lambda}{\partial s}\right) \qquad (3)$$

Typical $\Theta(s)$ curves are sketched in Figure 2. The width of the transition region is of the order $\sigma^{1/2}$, which clearly demonstrates the effect in the degree of cooperativity of the transition.

We now turn to a discussion of the intermolecular M-S coupling [3]. We assume, for the primitive model that we are considering here, that there exists an interaction between the helical domains on different molecules tending to line up along some preferred axis in a nematic configuration. In the spirit of a mean-field approximation, we write that each monomer that belongs to a helical segment experiences a dimensionless molecular field h:

$$h = c\Theta v \langle Q \rangle \qquad (4)$$

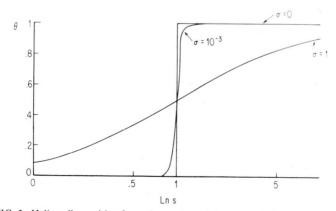

FIG. 2. Helix-coil transition for various values of the cooperativity parameter σ.

where c is the concentration of monomers, v is a phenomenological coupling constant with dimensions of volume, and

$$Q = \frac{1}{2}(3 \cos^2 \phi - 1) \tag{5}$$

is a quadrupole component whose thermodynamic average $\langle Q \rangle$ is the nematic order parameter. Here ϕ is the angle between the orientation of a given rigid segment and the mean alignment axis. The essence of the mean-field approximation is to suppose that the probability of a given segment having an orientation angle ϕ is proportional to $e^{hQ(\phi)}$. Then the activity coefficient s for each "helical" monomer will be replaced by se^{hQ}. In the calculation of the partition function [1, 2], every factor of s is renormalized to

$$\tilde{s} = \alpha s \tag{6}$$

$$\alpha = \frac{1}{2} \int_{-1}^{1} e^{hQ} \, d(\cos \phi) \tag{7}$$

With this definition of the renormalized activity coefficient, the free energy per monomer λ becomes $\tilde{\lambda}$, which has the same functional dependence on $\tilde{s}$ as $\lambda(s)$, i.e., $\tilde{\lambda} = \lambda(\tilde{s})$. The average nematic order parameter is

$$\langle Q \rangle = \alpha^{-1} \frac{\partial \alpha}{\partial h} \tag{8}$$

which leads to the self-consistency equation

$$h = \left(\frac{g\tilde{s}}{\alpha\tilde{\lambda}}\right) \left(\frac{\partial\tilde{\lambda}}{\partial\tilde{s}}\right) \left(\frac{\partial\alpha}{\partial h}\right) \tag{9}$$

which can be solved in terms of eqs. (1)–(8).

In this study we are content to consider the two limiting cases of eq. (9): (1) cooperative case $\sigma \rightarrow 0$ and (2) no intrinsic cooperativity $\sigma \rightarrow 1$. The intermediate case is under study.

Cooperative Limit $\sigma \rightarrow 0$

Here $\tilde{\lambda} = 1$ or $\tilde{s}$, whichever is larger. Thus, the transition to rigid molecules occurs at $\tilde{s} = s\alpha = 1$. Already we note that because $\alpha > 1$, the transition occurs for $s < 1$; i.e., the critical value of s, s^* is given by $s^* = \alpha^{-1}$, where α is the solution of the M-S equation, derived from eq. (9):

$$h = \left(\frac{g}{\alpha}\right) \frac{\partial\alpha}{\partial h} \tag{10}$$

where $g = cv$ is a dimensionless coupling constant. For $g < g_c = 4.4$, the only solution to eq. (10) is $\alpha = 1$; i.e., no nematic order and hence $s^* = 1$. On the other hand, as the polymer concentration increases to $g = g_c$, there exists a first-order jump in α to $\alpha_c = 1.5$, leading to $s_c = 0.67$. At this point, long-range nematic order develops with $\langle Q \rangle = 0.43$. As the coupling constant grows beyond g_c, α

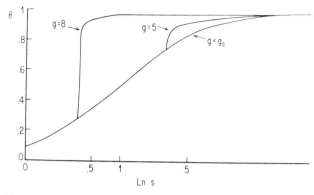

FIG. 3. The Ising model ($\sigma = 1$) with mean field intermolecular M-S interactions for various values of the coupling constant g.

increases approximately as $(e^h/3h)$, where $h = 1/2[g + (g(g - g_c))]^{1/2}$, with a corresponding decrease in s^*. The effect of the intermolecular coupling already becomes apparent in this limit. However, it is difficult to assess the change in order of the transition, since the isolated molecule helix-coil transition is arbitrarily sharp.

No Intrinsic Cooperativity $\sigma \to 1$

In this case the self-consistency equation is simply

$$h = \left[\frac{g\tilde{s}}{(1 + \tilde{s})} \right] \left(\alpha^{-1} \frac{d\alpha}{dh} \right) \tag{11}$$

In Figure 3, we sketch some numerical solutions of eq. (11). Note that if $g > g_c$, i.e., for sufficiently concentrated solutions, the helix-coil transitions seem to have an induced first-order character.

DISCUSSION

We have shown that if it is assumed that rigid rodlike segments of a macromolecule have a mutual coupling of quadrupolar symmetry and, furthermore, if the polymer concentration is sufficiently high, a first-order helix-coil mesomorphic transition may be obtained. The direction of future studies should concentrate on a fundamental investigation of the intermolecular interaction during a helix-coil transition in order to justify or negate the assumed model. It will probably be necessary to include both steric excluded volume interactions in a consistent manner as well as van der Waals attractions. More specifically the coil-coil interactions that have been omitted here must also be included.

REFERENCES

[1] N. S. Murthy, J. R. Knox, and E. T. Samulski, *J. Chem. Phys., 65,* 4835 (1976).
[2] T. M. Birshtein and O. B. Pititsyn, "Conformations of Macromolecules," Wiley-Interscience, New York, 1966, Chap. 9 and 10.
[3] P. G. de Gennes, "The Physics of Liquid Crystals," Clarendon Press, Oxford, 1975, Chap. 2.

KINETIC ASPECTS OF THE FORMATION OF THE ORDERED PHASE IN STIFF-CHAIN HELICAL POLYAMINO ACIDS

WILMER G. MILLER, LEE KOU, KOHJI TOHYAMA, and VINCENT VOLTAGGIO

Department of Chemistry, University of Minnesota, Minneapolis, Minnesota 55455

SYNOPSIS

The helical polyamino acids exhibit a two-component temperature-composition phase diagram that corresponds well in overall appearance to the Flory prediction for rigid, impenetrable rods. The phase diagram may be characterized as a narrow, typically 1–2%, biphasic region separating the disordered (isotropic) state from the ordered (liquid–crystalline) state, connected to a wide biphasic region in which nearly pure polymer is in equilibrium with nearly pure solvent. Initial studies have been undertaken to deduce the kinetics of formation of the ordered phase, both when the thermodynamically stable state is the ordered state, as well as when it is in an ordered-disordered biphasic state. The time course of the formation of the ordered phase at different temperatures is reminiscent of analogous studies on polymer crystallization, describable by a nucleation and growth mechanism. Formation of the ordered phase when the final state is the narow biphasic state appears also to follow a nucleation-growth mechanism. However, when the final state is the wide biphasic region, the morphology of the final state is that of an optically clear, self-supporting gel, even at very low concentrations (as low as 0.03 wt %). The reasons for and nature of the gel state are believed to be a kinetic phenomenon. The shape of the phase diagram is such that the temperature range of thermodynamic metastability is small and phase separation occurs in the thermodynamically unstable region. In the unstable region the zero activation-energy spinodal decomposition mechanism dominates the kinetics of phase separation, leading to bicontinuous interconnected phases that have little possibility or driving force to rearrange further. The polymer network structure is proposed to be a bulk thermodynamic phase, characterized by bundles or sheets of polymer between branch points, rather than individual chains as in the usual polymer network. Rheological, electron microscope, and light-scattering data are presented that support this view. The possibility of other stiff-chain polymers forming a network structure for similar reasons is discussed.

INTRODUCTON

The existence, organization, and stability of an ordered phase of stiff-chain helical polyamino acids have been the object of numerous studies over the past 25 years. Kinetic aspects of the formation of the ordered phase have received little attention. One of the reasons for this is that an understanding of the kinetics is dependent on a knowledge of the thermodynamic phase boundaries, which for most polypeptide–solvent pairs is known in only a limited way. The phase equilibria for poly(γ-benzyl-α,L-glutamate) (PBLG) and for poly(ϵ-carbo-

Journal of Polymer Science: Polymer Symposium 65, 91–106 (1978)
© 1978 John Wiley & Sons, Inc.   0360-8905/78/0065-0091$01.00

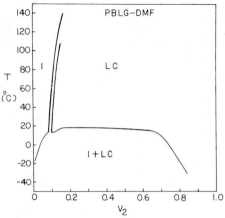

FIG. 1. Temperature-composition phase diagram for PBLG (310,000) in DMF [1–3]. The phases are disordered (I) and ordered (LC).

benzoxy-α,L-lysine) (PCBL) have been described in most detail [1–4]. We have, therefore, concentrated our studies on these polymers, using N,N-dimethyl-formamide (DMF), benzene, and toluene as solvents.

The temperature-composition phase diagrams for PBLG and for PCBL in DMF [1–4] are shown in Figures 1 and 2, respectively. The theoretical phase diagram of Flory for rigid, impenetrable rods [5] is shown in Figure 3 for comparison. The major predictions of the Flory theory are realized, and the detailed differences are primarily a result of the experimental rods (helices) being neither completely rigid nor solvent-impermeable [2]. The phase equilibria can be characterized as consisting of a narrow chimney in which there is only a small

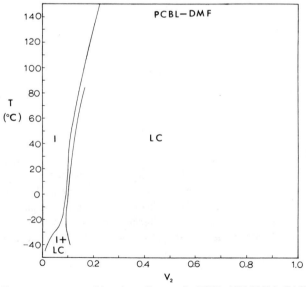

FIG. 2. Temperature-composition phase diagram for PCBL (690,000) in DMF [2–4].

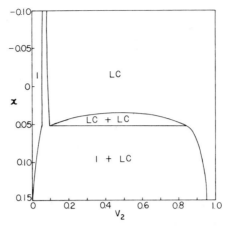

FIG. 3. Phase equilibria for rigid, impenetrable rods [5] with axial ratio 150.

difference in composition between the ordered and disordered phases, connected to a wide biphasic region in which the polymer and solvent mix only slightly. The solubilization and molecular dispersal of PBLG and PCBL as rods in DMF is dominated by the free energy of mixing the flexible, partially polar polymer side chains with the polar DMF [6, 7]. As the polarity of the solvent is reduced, the polymer has a tendency to associate considerably, even in dilute solution [8]. This, however, does not affect significantly the phase boundaries of the chimney or narrow biphasic region. For example, with PBLG at room temperature the liquid–crystal phase first appears at about 8.5 vol % polymer in DMF [1–3], a polar, nonassociating solvent [8]; at 8 ± 0.5 vol % in methylene chloride [9], a less polar and weakly associating solvent; and at 7 ± 0.5 vol % in dioxane [9], a nonpolar solvent in which the polymer strongly associates even at low concentrations [8]. Under the same conditions the single-phase liquid–crystal region appears at about 11.5 vol % in DMF [1–3], at 12–15 vol % in methylene chloride [9], and at 9–11 vol % in dioxane [9]. In addition to association the main effect of decreasing the solvent polarity is to raise the temperature at which the narrow and wide biphasic regions intersect. Thus, studies in any region of the phase diagram may be made by varying the temperature or by an appropriate adjustment of the solvent to bring a particular region to a temperature of experimental accessibility.

EXPERIMENTAL

Poly(benzyl glutamate) of 310,000 weight-average molecular weight M_w was obtained from Pilot Chemicals, and 150,000 molecular weight from Miles Laboratories. Poly(carbobenzoxylysine) of 1.5×10^6 wt av mol wt was obtained by polymerizing carbobenzoxylysine N-carboxyanhydride in dioxane using triethylene amine as initiator. Reagent grade solvents were used as received.

The formation of the ordered phase after quick quenching from the isotropic phase was followed by measuring the light coming through crossed polars at a fixed wavelength using a spectrophotometer fitted with polars on either side of the sample cell. The phase change was observed also by following the disappearance of the high-resolution phenyl proton nuclear magnetic resonance signal using a Varian XL-100 spectrometer. Previous studies [4] have shown that the phenyl proton resonance can be used as a quantitative measure of the presence of the ordered phase.

The structure of the gel phase was explored using rheological, electron microscope, and light-scattering measurements. Dynamic shear moduli were measured with a Rheometrics rheometer fitted with eccentric, rotatable, parallel disks. The gel was formed on the lower disk and later put into contact with the upper disk. Moduli were determined both in the static offset and in the eccentric, rotating disk mode.

Electron micrograph replicas were prepared by quick-freezing the gel, followed by freeze fracturing, etching, and platinum shadowing at $-100°$ in a Bolgers 360 apparatus. The replicas were examined with a Zeiss EM9 or a RCA EMV 4C transmission microscope.

FORMATION OF THE SINGLE-PHASE ORDERED STATE FROM THE ISOTROPIC STATE

The shift of the narrow biphasic region to higher concentrations with increasing temperature, a result of increased rod flexibility [2], makes possible a convenient approach to studying the kinetics of the ordered phase formation. An isotropic solution can be temperature-jumped across the biphasic region into the region in which the ordered phase is stable. When an isotropic solution being maintained, for example, at 80° is suddenly cooled to a temperature at which the ordered phase is the thermodynamically stable state, the time course of the appearance of the ordered phase depends dramatically on the temperature. This is shown in Figures 4 and 5, in which the data have been scaled and plotted as the fractional change. At the point at which increases in transmitted light per unit time become very small, the sample appears cloudy. Viewed between crossed polars in a microscope, the light is found to be depolarized, but the bright multicolored appearance and the periodicity lines typical of aged, ordered phases are not yet present. If, after the change in intensity has leveled off, the temperature is again raised to 80°, the light intensity returns to its original value within a few minutes. The proton nmr of the benzyl protons has been shown to give the fraction of rods in the ordered state nearly quantitatively [4]. In the ordered state, the benzyl protons make no contribution to a high-resolution spectrum, in contrast to their contribution in the disordered state. In Figure 5 is shown the time dependence of the benzyl proton nmr signal. The disappearance of the disordered phase coincides with the leveling off of the light transmitted through crossed polars. The nmr together with the optical data is consistent with the idea that we are observing random nucleation and growth of the ordered phase. The time at which light transmission approaches a pseudosteady value corresponds to the point at which the polymer has been converted quantitatively to the ordered phase, as depicted in Figure 6. The cloudy appearance of the sample at

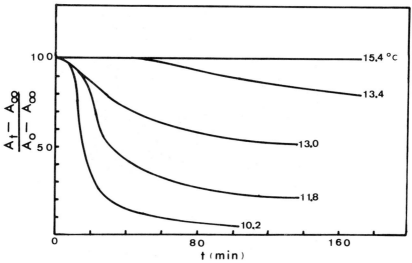

FIG. 4. Appearance of the ordered phase as measured by light passing crossed polars. The sample is 13.3 wt % PCBL (1.5×10^6) in DMF. The initial temperature was 80°. The temperature of ordered-phase formation was as indicated.

this point is indicative of multiple, randomly oriented domains. If the sample is further aged, some domains grow at the expense of others until a single domain may extend over macroscopic dimensions, as evidenced by the light microscope view in Figure 6. The time scale for this annealing process is orders of magnitude

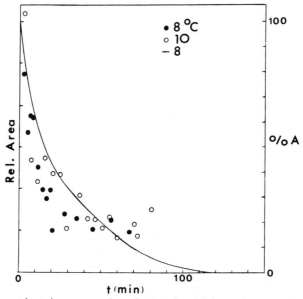

FIG. 5. Apparent benzyl proton nmr intensity (●,○) and light passing crossed polars (−) as a function of time. Concentration is slightly higher than in Figure 4. Large uncertainty in proton intensity is a result of broadened resonance due to the very high molecular weight.

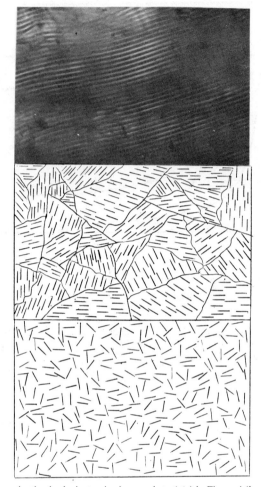

FIG. 6. State of organization in the isotropic phase and at $\tau(A_0)$ in Figure 4 (lower); in the ordered phase after light transmission approaches a pseudoconstant value at $\tau(A_\infty)$ (middle); in the ordered phase at a much later time when an ordered domain extends over macroscopic dimensions as revealed by the periodicity lines observed through a polarizing microscope (upper).

longer than that for phase formation observed in Figures 4 and 5. During the annealing further changes in light transmission occur; hence, the A_∞ values are, in fact, pseudovalues.

The time course of the ordered-phase formation as represented in Figures 4 and 5 is reminiscent of the kinetics of crystallization of polymers from the melt or from solution [11]. Unlike the mobility problems that arise in polymer crystallization studies, the ordered phase here is quite fluid. The kinetic analysis should be simplified, as mobility does not change when the completely ordered state is approached. If the data are fitted to the Avrami equation [11], as is customary with polymer crystallization data, the power dependence of the time (the Avrami n) varies with the temperature of phase formation. The analysis of the kinetics is not trivial. The temperature dependence of the nucleation is complex, as the enthalpy of forming the ordered phase is positive, but itself has a considerable temperature dependence because of the superimposed effect of rod flexibility [2]. By a combination of optical, volume, and nmr measurements we are currently working out the details of the ordered-phase formation. It seems clear, however, that the kinetic mechanism of the ordered-phase formation is nucleation and growth.

ORDERED-PHASE FORMATION WHEN THE FINAL STATE IS BIPHASIC

Experimental Observations

When the temperature is adjusted so that the final state is in the narrow biphasic or chimney region of the phase diagram, ordered spherulites observable by light microscopy appear in the disordered phase [9, 10]. On annealing, the spherulites tend to settle, as the ordered phase is slightly more dense. The density difference and the interfacial tension are sufficient to cause a menicus to form in the PCBL–DMF system; in the PBLG–DMF system no meniscus forms [2]. Although the time scale becomes long as the driving force to form the ordered phase becomes small, the appearance of spherulites suggests a nucleation mechanism.

When an isotropic, biphasic, or ordered solution is brought into the wide biphasic region of the phase diagram, the solutions are always observed to form a transparent, mechanically self-supporting gel [1, 2]. Light coming through crossed polars is depolarized, but the bright colors and the periodicity lines observed in the single ordered-phase region never develop; instead, with white light a uniform, pale yellow field is seen. Gelation has been observed previously [8] but was not related to phase formation. The gel formation is concentration- and temperature-dependent and completely reversible. Gelation occurs in the PBLG–DMF system 10–15° above the temperature at which light is first observed coming through crossed polars, which we have designated previously as the phase boundary separating the isotropic from the biphasic region. Below about a 1% solution the depolarized light intensity is too small to detect visually. However, as is shown in Figure 7, in the measurable concentration range the temperature dependence has a simple functional form, given by $v_2 = 1 \times 10^{-6}$ exp $(-4650/T)$. If this functional form were to prevail, it would correspond to a volume fraction of 0.0003 at $-60°$, the freezing point of DMF. Experimentally, a 0.05 vol % solution gels at about $-40°$. From Figure 7, v_2 is 0.0005 at $-55°$. The extrapolation and relationship to gelation thus appear to hold even in very dilute solutions.

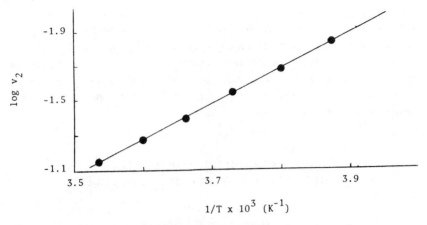

FIG. 7. Temperature dependence of the isotropic-biphasic boundary taken as the temperature at which light is first observed passing crossed polars.

A Kinetic Basis for Network Formation

In polymer terminology a mechanically self-supporting gel implies a three-dimensional network structure. In our system there is no possibility of covalent crosslinks, of noncovalent crosslinks such as hydrogen bonds, or of entrapment of polymer chains between crystallites to give pseudocrosslinks. But yet we are able to form self-supporting gels with solutions as dilute as 0.03 wt % polymer, and the gelation seemed related to crossing the phase boundary into the wide biphasic region. After considerable scrutiny of the literature, we concluded that the gelation might be a kinetic phenomenon related to the kinetic mechanism of phase formation. In particular, we thought that the Cahn-Hillard spinodal decomposition mechanism [12, 13] for phase formation in the thermodynamic region unstable with respect to fluctuations ($\partial^2 G / \partial c^2 < 0$) might be applicable. Unlike nucleation, this mechanism requires no activation energy, with phase separation occurring by a continuous unmixing characterized by a fluctuation vector. Of particular interest is the prediction that early in the phase formation bicontinuous phases may form, in which each phase is continuous in the other phase [14], whereas in the usual nucleation mechanism the nucleated phase is not continuous. At later times the system may ripen, because it is driven by interfacial tension to minimize the surface area, which leads eventually to phases no longer continuous in the other phase. Several polymer systems have been investigated from the viewpoint of the spinodal mechansim [15–23]. Of special interest is the work of McMaster [23] in which the existence of bicontinuous phases early in the phase separation is more clearly demonstrated than in other studies. Ultimately, however, connectivity is lost.

In applying the spinodal decomposition mechanism to explain gelation in our system, there are several points to be raised and discussed. The flatness of the phase boundary around the critical temperature (Fig. 1) ensures that the spinodal lies very close to the binodal (phase boundary) over a considerable composition range. This minimizes the temperature range of metastability in the concentration extremes. It may, thus, be relatively easy to pass through the metastable into the unstable region, in contrast to a more typical phase diagram in which the phase boundary is not so flat around the critical point.

Even if our network structure is formed as a result of bicontinuous phase formation early in the phase separation process, why does it remain bicontinuous instead of ripening into two bulk phases? A rationale for this can be devised from the phase diagram. On entering the wide biphasic region, the polymer-rich phase is effectively bulk polymer, particularly with respect to its rheological properties. Bicontinuous polymer phases studied previously have been observed to age by viscous flow in each phase [23]. In our case the polymer continuous phase should be essentially fiberlike, and significant viscous flow is not expected. An alternate aging mechanism is possible, by which polymer is transported across the solvent—continuous phase. Two factors would minimize this as a major aging mechanism: a low driving force due to low interfacial tension, or a low solubility of the polymer in the solvent continuous phase. The latter condition is satisfied only a few degrees below the critical temperature. The interfacial tension is not known.

A final point to be considered is the conclusion of Cahn that bicontinuous phases would not be formed unless the volume fraction of the minor phase exceeds about 0.15 [14]. Below this, isolated second-phase particles rather than an interconnected or network structure would form. The resultant morphology of the polymer phase is predicted to be similar to nucleation and growth, except that the particles have a tendency to be a fixed distance apart, controlled by the dominant spinodal fluctuation vector. The calculations of Cahn were made for a two-component system of molecules of equal size, assuming isotropic behavior. In our system the ratio of molar volumes of polymer and solvent is typically 1000 to 3000, the polymer has an axial ratio of 50 to 150, and, because of the molecular asymmetry, the growth of the polymer phase may be very anisotropic. The effect of these asymmetries on the phase connectivity is not clear. At the minimum the spinodal decomposition mechanism directs spatially where the polymer phase will form. At later times, particularly if the growth is anisotropic, the polymer phase that is being formed becomes interconnected and retains that morphology.

It is, thus, our view that the gel may not be a "classical" network structure with single polymer chains spanning branchpoints. Instead, the network may be a thermodynamic bulk phase composed of bundles or sheets of aligned rods spanning branchpoints. Its formation is controlled by the kinetic mechanism for phase separation when the system is in a thermodynamically unstable state.

EXPERIMENTAL STUDIES ON THE GEL STATE

Our initial efforts have been put into characterizing the gel state rather than in making direct kinetic studies. With PBLG in DMF the gel forms only below room temperature (Fig. 1). It is not possible using an air bath and a volatile solvent to make rheological measurements at subambient temperatures. Consequently, the solvent was changed to toluene in which concentrations higher than 0.1 wt % polymer give gels at room temperature. In Figure 8 are shown rheological data for a 1 wt % PBLC solution in toluene at about 25°. This sample is 35° below the gelation temperature. Static measurements gave very similar values for G'. The data in Figure 8 are for a cylinder of gel 5.0 cm in diameter and 0.42 cm thick. The frequency range is very limited, as the gel tends to lose contact with the upper disk at higher frequencies. Thinner gel cylinders gave similar but less precise results. The concentration dependence of the modulus was explored roughly; it did not appear to be strongly concentration-dependent. By comparison gelatin gels have a c^2 dependence [24], and polyvinyl chloride gels an even higher dependence [25]. Over the limited frequency range studied, the moduli showed little frequency dependence. Few moduli have been measured on gels containing 1% or less polymer. The storage modulus of 10^4 dyn cm^{-2} seems to be higher than is generally encountered in such dilute gels. The loss tangent is quite low. The data are consistent with what is expected for an elastic network, with a concentration dependence much less than is typically found in normal network-forming systems.

When the temperature of a solution is lowered across the phase boundary to some point inside the wide biphasic region, the time required for the system to set to a self-supporting network ranges from minutes to hours, depending on the concentration and degree of undercooling. If, during the period leading to iso-thermal gelation, part of the solution is subjected to a shear force, e.g., by the action of a magnetic stirring bar rotating so that the bottom part of the sample is stirred while the upper part remains virtually undisturbed [1], a quite repeatable phenomena occurs. In the upper, unstirred part a clear gel forms. In the stirred part, a gel does not form. Instead, the solution is very fluid and cloudy. These two morphologies coexist in contact for extended periods (>1 year) with no tendency for one to grow at the expense of the others. We interpret the cloudiness as light scattered from fragments of the network dispersed in the solvent. The fact that the fragments remain in contact with the gel without either

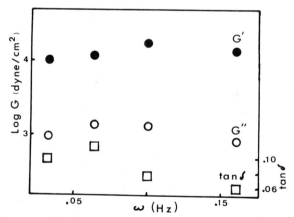

FIG. 8. Dynamic storage G' and loss G'' modulus and loss tangent (tan δ) as a function of frequency at 25° for 1 wt % PBLG (150,000) in toluene.

system undergoing any apparent change suggests that there is an exceedingly low driving force for minimizing the surface area between the polymer-rich and solvent-rich phases, once the gel has formed.

An attempt was made to visualize the polymer morphology. A gel sample was prepared, quick-frozen, fractured, etched for 1 or 2 min, shaded with platinum, backed, and stripped of gel by soaking in DMF. The mobility of the polymer molecules is so low that they cannot diffuse during the quick freezing of a few hundredths of a milliliter of gel. Some electron micrographs are shown in Figures 9 and 10. Conditions are as indicated. Single molecules, being only about 20 Å across, are not visible. An interconnected polymer phase does exist and has a structure that seems compatible with the spinodal decomposition mechanism.

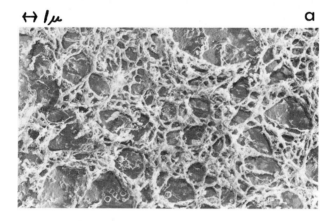

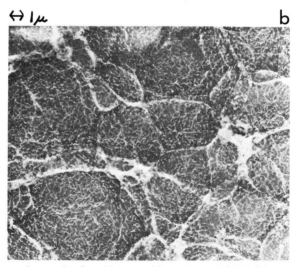

FIG. 9. Electron micrographs of a gel 0.2 wt % (a) or 0.5 wt % (b) PBLG (150,000) in toluene.

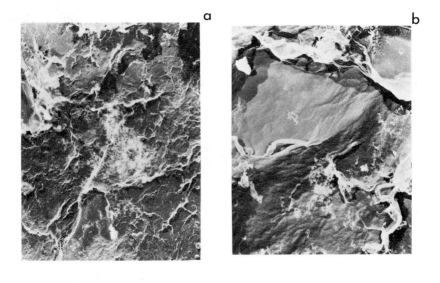

FIG. 10. Electron micrographs of a gel 0.2 wt % PBLG (150,000) in toluene (a, b) and in a similar gel partially melted before freezing (c).

In Figure 10(c) the sample had warmed slowly, and the solution was frozen in the partially reversed gel state. In this instance, the micrograph appears quite different, showing bundles of rods that have been cut across in the fracture and visualized by etching. Although the distinction between sample and artifact can be difficult in electron microscopy, the technique of etching and shadowing a freshly fractured surface in the interior of the frozen gel seemed to us to offer the best approach to visualizing the polymer morphology in the gel state. The morphology may be strongly dependent on the concentration and temperature of gel formation. This is discussed below.

A further attempt to characterize the gel was made by quasielastic light scattering. Prins [20–22] and Benedict [26] have shown that the autocorrelation function can be related to the elastic properties of the gel network. In crosslinked polyacrylamide gels, Benedict [26] has shown that the autocorrelation function can be fitted to a single exponential. The decay constant depends on the shear modulus and friction factor of the gel. The values obtained by light scattering compare favorably with those obtained by macroscopic measurement. Similar measurements [27] on PBLG solutions and gels are shown in Figure 11. Polyacrylamide gels behaved similarly to those of Benedict. PBLG solutions above the gel temperature showed an exponential decay. Gels 10° or so below the gel temperature behaved as in Figure 11(b). No correlation function was observable. At temperatures near the gel temperature the autocorrelation function had a fluctuating decay that persisted for days after the gel was formed; as, for example, in Figure 11(c) for a 1.2 wt % PBLG in benzene gel. Similar results were obtained in DMF, benzene, and toluene. The fluctuating autocorrelation function appears similar to that observed by Prins [22] and attributed to nonequilibrium mass flow. Our inability to observe the decay in the autocorrelation function in considerably undercooled PBLG gels is likely due to the very high elastic modulus in the very dilute gels and perhaps to a small friction factor. The light-scattering data are considerably different from those for typical low-

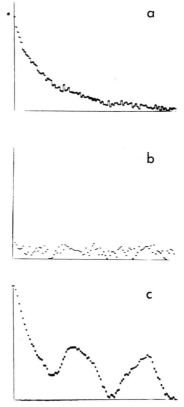

FIG. 11. Decay in autocorrelation function (a) in polyacrylamide gels or PBLG solutions; (b) in PBLG gels at temperature well below the gel temperature; (c) in PBLG gels near the gel temperature.

concentration gel networks composed of single chains spanning network branchpoints and seem consistent with the bundle and sheet structures proposed to form from the spinodal decomposition mechanism. A fuller accounting of the light-scattering data is given elsewhere [28].

DISCUSSION

We have initiated studies to elucidate the kinetics of forming the ordered phase in the stiff-chain helical polyamino acids. When the final state was the ordered phase, we assumed that the kinetics could be described by a nucleation and growth mechanism with many similarities to the kinetics of polymer crystallization. What we have shown here is that the time scale is such as to allow the development of the ordered phase to be followed by optical, volume, and nmr techniques. We have shown, as presented in Figures 4–6, that the kinetic process can be divided into two time scales: the conversion of the randomly oriented rods to a random array of locally oriented rod domains, followed by growth of some domains at the expense of others. A detailed analysis of the conversion to the locally oriented domains is under investigation.

What is of particular interest in our efforts is the finding that the kinetic mechanism of phase formation when the final state is biphasic appears to depend on whether we are in the narrow or wide biphasic region. The existence of the narrow and wide biphasic regions with a conversion from one to the other over a narrow temperature range is a feature peculiar to stiff-chain polymers, as Flory predicted. The narrow biphasic region, a result of molecular asymmetry, is little affected by polymer–solvent interactions. The wide biphasic region, on the other hand, is controlled by polymer–solvent interactions as a small change in solvent power has a catastrophic effect on polymer–solvent mixing. It is this sudden unmixing which leads, we believe, to a change in the mechanism of ordered phase formation from nucleation and growth to a zero activation energy, spinodal decomposition. We were forced to this mechanism in an attempt to explain the appearance of polymer networks on entering the wide biphasic region. Although this mechanism has received attention in the polymer literature, it is the first time, to our knowledge, that it has been proposed in explaining the formation of a stable network morphology. In conceptual terms the weakest point in its application to our network structures is the formation of network at exceedingly low polymer concentrations, in which the volume fraction of the polymer phase is less than that predicted to lead to even initial-stage connectivity. If, under further scrutiny, this mechanism is viable, asymmetric growth on top of the spinodally determined spatial placement will most likely be the answer.

Of special importance to the resultant morphology of the spinodal decomposition mechanism is the fluctuation vector or fluctuation wavelength. It depends on the quenching temperature relative to the spinodal temperature, and on the range of molecular interaction, generally taken as the radius of gyration [15–17]. It thus depends on temperature, molecular weight, and polymer concentration. The resultant morphology, in particular, the mean distance between polymer "bundles" as well as "bundle" thickness, should change considerably over a few degrees change in temperature or a small change in concentration. It might even depend on sample size, as a thick sample when quenched attains temperature equilibrium more slowly than a thin sample. In most of the work presented here the spacing should be of the order of microns and the "bundle" width typically an order of magnitude lower. This is reasonably consistent with at least some of the electron microscope results.

It should be noted that if the spinodal decomposition mechanism is applicable, the low concentration "phase boundary" as determined by light microscopy (Figs. 7 and 1) is meaningless. As we have noted, the gel points are $10°-15°$ higher than the microscope values but parallel them. It, thus, seems more reasonable to classify the gel points as the spinodal, with the binodal or regular phase boundary lying at somewhat higher temperatures. An attempt to determine the spinodal and binodal lines is discussed elsewhere [29].

A final point to discuss is the applicability of these results to other stiff-chain polymers. In principle, the Flory theory should be applicable to any stiff-chain polymer. Other than the helical polyamino acids few synthetic, high-molecular-weight, stiff-chain polymers have been made until recently; so little is known in detail about phase equilibria. We are not aware of gel phases being reported. But by analogy with our own work, they could be missed easily. With many solvents, the full phase diagram is not observed over the liquid range of the solvent. With some solvents only the narrow biphasic region is observed down to the solvent freezing point; with others only the wide biphasic region is observed up to the boiling point. Only in the accidental case in which the transition occurs during the liquid range of the solvent would we anticipate network formation. In the biological world there are many examples of stiff-chain polymers. Some exist always in this form, others may be random coils under some conditions, and others form by polymerization of globular macromolecules, for example, sickle-cell hemoglobin. The biological literature makes reference to many rodlike macromolecules that gel, for example, DNA, gelatin, sickle-cell hemoglobin, myosin, actin, tobacco mosaic virus, and fibrin. Although it is clear that some of these clearly do not gel by the spinodal decomposition mechanism, none have been investigated from this viewpoint.

We wish to acknowledge the support of the USPHS (GM16922). We wish to thank Professor C. Macosko for use of the Rheometrics rheometer, and Professor V. Bloomfield and George Beren for light-scattering measurements [28].

REFERENCES

[1] E. L. Wee and W. G. Miller, *J. Phys. Chem., 75,* 1446 (1971).
[2] W. G. Miller, L. L. Wu, E. L. Wee, G. L. Santee, J. H. Rai, and K. D. Goebel, *Pure Appl. Chem., 38,* 37 (1974).
[3] W. G. Miller, J. H. Rai, and E. L. Wee in "Liquid Crystals and Ordered Fluids," J. F. Johnson and R. S. Porter, Eds., Vol. 2, Plenum, New York, 1974, p. 243.
[4] G. L. Santee, M.S. Thesis, University of Minnesota, 1972.
[5] P. J. Flory, *Proc. R. Soc. London, A234,* 73 (1956).
[6] J. H. Rai and W. G. Miller, *Macromolecules, 5,* 45 (1972).
[7] J. H. Rai and W. G. Miller, *Macromolecules, 6,* 257, (1973).
[8] P. Doty, J. H. Bradbury, and A. M. Holtzer, *J. Am. Chem. Soc., 76,* 947 (1956).

[9]   C. Robinson, J. C. Ward, and R. B. Beevers, *Discuss. Faraday Soc., 25,* 29 (1958).

[10] C. Robinson, *Trans. Faraday Soc., 52,* 571 (1956).

[11] L. Mandelkern, "Crystallization of Polymers," McGraw-Hill, New York, 1964.

[12] J. W. Cahn, *Trans. Metall. Soc. AIME, 242,* 166 (1968).

[13] J. E. Hillard in "Phase Transformation," H. I. Aronson, Ed., American Society for Metals, Metals Park, Ohio, 1970.

[14] J. W. Cahn, *J. Chem. Phys., 42,* 93 (1965).

[15] J. J. van Aartsen, *Eur. Polym. J., 6,* 919 (1970).

[16] J. J. van Aartsen and L. A. Smolders, *Eur. Polym. J., 6,* 1105 (1970).

[17] C. A. Smolders, J. J. van Aartsen, and A. Steenbergen, *Kolloid Z. Z. Polym., 243,* 14 (1971).

[18] P. T. van Emmerik and C. A. Smolders, *J. Polym. Sci. Part C, 38,* 73 (1972).

[19] P. T. van Emmerik and C. A. Smolders, *Eur. Polym. J., 9,* 309 (1973).

[20] E. Pines and W. Prins, *Macromolecules, 6,* 888 (1973).

[21] G. T. Feke and W. Prins, *Macromolecules, 7,* 527 (1974).

[22] K. L. Wun, G. T. Feke. and W. Prins, *Faraday Discuss. Chem. Soc., 57,* 146 (1974).

[23] L. P. McMaster in "Copolymers, Polyblends and Composites," N. A. J. Platzer, Ed., *Adv. Chem. Ser., 142,* American Chemical Society, Washington, D.C., 1975.

[24] A. Veis, "The Macromolecular Chemistry of Gelatin," Academic, New York, 1964.

[25] J. D. Ferry, "Viscoelastic Properties of Polymers," Wiley, New York, 1970.

[26] T. Tanaka, L. O. Hocker, and G. B. Benedek, *J. Chem. Phys., 59,* 5151 (1973).

[27] J. Aksiyote-Benbasat and V. A. Bloomfield, *J. Mol. Biol., 95,* 335 (1975).

[28] K. Tohyama, G. Berens, V. Bloomfield, and W. G. Miller, to appear.

[29] E. Wee, L. Kou and W. G. Miller, ACS Meeting, Chicago, August 29–September 2, 1977.

PROPERTIES OF THE LADDER POLYMER CIS-SYNDIOTACTIC POLY(PHENYLSILSESQUIOXANE) IN SOLUTION

T. E. HELMINIAK* and G. C. BERRY

Department of Chemistry, Carnegie-Mellon University, Pittsburgh, Pennsylvania 15213

SYNOPSIS

Studies on the properties of the double-chain ladder polymer of *cis*-syndiotactic poly(phenylsilsesquioxane) (PPSQ) in dilute and concentrated solutions are reported. Data on the intrinsic viscosity $[\eta]$, the second virial coefficient A_2, and the mean-square radius of gyration $\langle s^2 \rangle$ as functions of molecular weight M, temperature, and solvent lead to the conclusion that the conformation of PPSQ can be modeled with a wormlike chain model with a persistence length ρ of 74 Å. These data cannot be used to assess the extent of intramolecular inflexibility, e.g., resistance to rotational isomerization, of PPSQ. For concentrated solutions ($0.05 \leq c \leq 0.20$ g/ml) the limiting viscosity η_0 at zero shear rate is proportional to cM for $cM < \rho_2 M_c$ and to $(cM)^{3.4}$ for $cM \geq \rho_2 M_c$, where $M_c = 9120$. The product $X_c = (\langle s^2{}_0 \rangle / M) \times \rho_2 M_c / m_a$ is equal to 356×10^{-17} for PPSQ, close to the value 400×10^{-17} typical of single-chain polymers. The reduced flow curve η_κ / η_0 versus $\tau_c \kappa$ for the viscosity η_κ as a function of shear rate κ is not far different from that for single-chain polymers—here, the time constant τ_c is equal to $\eta_0 R_0$ with R_0 the limits of the recovery function R_κ at small κ (R_0 is often denoted $J_e{}^0$). The values of τ_c are intermediate to τ_c for a flexible single-chain polymer and the relaxation time τ_R for rotatory diffusion of a rigid coil. Plots of R_κ versus κ superpose with plots of the transient recovery $R_0(\theta)$ with $\theta = \kappa^{-1}$ over the limited range $\tau_c \kappa < 1$ where θ is the recovery time ($R_0(\theta)$ goes to R_0 for large θ), similar to behavior found with solutions of single-chain polymers.

INTRODUCTION

Since the first synthesis [1] in 1960, the "ladder" polymer *cis*-syndiotactic poly(phenylsilsesquioxane) (PPSQ) has been the subject of several studies directed toward the elucidation of the rigidity of the linear double-chain backbone:

PPSQ

* Present address: Polymer Branch, Air Force Materials Laboratory, Wright-Patterson Air Force Base, Ohio 45433

Journal of Polymer Science: Polymer Symposium 65, 107–123 (1978)
© 1978 John Wiley & Sons, Inc.   0360-8905/78/0065-0107$01.00

Much of this work has been reviewed by Tsvetkov [2-4], who, indeed, has contributed many of the papers on the subject. These investigations have principally dealt with properties of dilute solutions, e.g., light scattering, viscometry, sedimentation, flow, electric birefringence, dielectric dispersion, and a few others. These studies suggest that PPSQ is a rigid, or inflexible, coil with a persistence length ρ of about 70 Å. In this study we present results on some rheological properties of moderately concentrated solutions (5-20 wt %) of PPSQ together with new light-scattering, viscometric, and exclusion chromatography (GPC) data on dilute solutions of PPSQ in a variety of solvents, including a theta solvent (ethylene dichloride at 50.5°). These data are analyzed along with osmotic pressure and viscometry data published previously by one of us [5].

It should be remarked at the outset that the term *inflexible chain* has been used to mean different things by different investigators. Here, the term is used in the sense that rotational isomerization of the chain backbone is prohibited, so that changes in the skeletal conformation must involve the bending and stretching of bonds (here primary valence bonds). This is to be distinguished from a chain configuration that might lead to a highly nongaussian segment density but for which intramolecular rotational isomerization is still possible. For example, PPSQ will be inflexible in the former sense if the intramolecular cyclization leading to the ladder structure is complete. If this cyclization is only partially complete, it might be regarded as a flexible chain by the criterion given, even though the gaussian chain would be inadequate as a model for predicting its properties, since the molecule will be far less coiled than is presumed with such a model. Tsvetkov has distinguished between these two properties by the use of the terms *kinetic flexibility* and *equilibrium flexibility*.

Finally, it should be mentioned that the configuration of PPSQ is known to depend markedly on the method of preparation [1]. The linear ladder chain structure depicted above is expected only under certain reaction conditions. In other circumstances, although the polymer has the formula $(C_6H_5SiO_{3.5})_x$, other modes of cyclization or even branched-chain configurations can result, the latter leading to gelation in some cases. The polymers used in this study were prepared under conditions giving the ladder configuration.

EXPERIMENTAL

Materials

Two samples of PPSQ were kindly furnished by Dr. J. F. Brown, of the General Electric Research Laboratories, Schenectady, New York. The fractionation of these is described elsewhere [5]. Briefly, the polymers were freed from possible contamination by up to 0.1% KOH by neutralization with acetic anhydride in benzene solution (9.7% polymer). After stirring overnight, the polymer was precipitated with methanol, filtered, washed, and dried to constant weight under vacuum at a maximum temperature of 100°C. The two samples, designated hereafter as A and B, were each fractionated by the standard coacervation technique using benzene and methanol as the solvent and nonsolvent, respectively, at 30°. Fractions were recovered as an opaque gelatinous product. Samples A (56 g) and B (101 g) were fractionated to give 6 and 10 fractions, respectively.

Reagent grade solvents used in solution studies were distilled under vacuum and stored under nitrogen before use.

Intrinsic Viscosity

Measurements were made using Cannon-Ubbelohde dilution viscometers (No. 25 or 50) having negligible kinetic energy corrections. Concentrations were adjusted to give relative viscosities η_{rel} in the range $1.05 < \eta_{rel} < 1.8$ for the four dilutions used with each sample. Data were plotted according to the usual relations

$$\frac{\eta_{sp}}{c} = [\eta] + k'[\eta]^2 c + k''[\eta]^3 c^2 + \cdots \tag{1}$$

$$\frac{\ln \eta_{rel}}{c} = [\eta] - \left(\frac{1}{2} - k'\right)[\eta]^2 c + \left(\frac{1}{3} - k' + k''\right)[\eta]^3 c^2 + \cdots \tag{2}$$

to determine the intrinsic viscosity $[\eta]$ and the Huggins constant k', with $\eta_{sp} = \eta_{rel} - 1$. Cannon variable-shear dilution viscometers providing for an eightfold change in the shear rate were used to assess the possible dependence of $[\eta]$ on shear rate with polymers A-1 and A-5 in benzene. No effect was found over the accessible shear rate range 30–900 sec^{-1} [5].

Light Scattering and Refractometry

A light-scattering photometer described in detail elsewhere [6] was used to determine the weight-average molecular weight M_w and the light-scattering averaged mean-square radius of gyration $\langle s^2 \rangle_{LS}$ and second virial coefficient A_2^{LS}. Data plotted at each concentration c according to the usual relation

$$\frac{Kc}{R_\theta} = \frac{Kc}{R_0} + \frac{1}{3}\frac{\langle s^2 \rangle_{LS}}{M_w} h^2 + O(h^4) \tag{3}$$

were extrapolated to give the intercept K_c/R_0 at zero scattering angle. Here, K is an optical constant proportional to the square of the refractive index increment dn/dc, and $h = (4\Pi n/\lambda_0) \sin \theta/2$ with θ the scattering angle, λ_0 the wavelength of light in vacuo, and n the refractive index. Plots of Kc/R_θ versus h^2 were found to be linear with a slope independent of c for a given sample. The extrapolated data K_c/R_0 were plotted according to the relation

$$\left(\frac{Kc}{R_0}\right)^{1/2} = \frac{1}{(M_w)^{1/2}}(1 + \Gamma_2^{LS} c + \cdots) \tag{4}$$

to determine M_w and $\Gamma_2^{LS} = A_2^{LS} M_w$. Vertically polarized light was used with λ_0 4358 and 6328 Å for studies with solutions in chloroform and tetrahydrofuran, respectively.

A differential refractometer described elsewhere [6] was used to determine dn/dc for solutions of PPSQ in chloroform and tetrahydrofuran. The results, 0.119 and 0.111 ml/g for λ_0 equal to 4358 and 5461 Å, respectively, for chloroform at 20° and 0.133 ml/g at 6328 Å for tetrahydrofuran at 20° are in good accord with data on solutions in benzene [7] (0.06 ml/g) and bromoform [8] (0.005 ml/g) when fitted by the Dale-Gladstone relation

$$\frac{dn}{dc} = v_2(n_2 - n_1) \tag{5}$$

with the specific volume of polymer $v_2 = 0.73$ ml/g [7]. Here, n_2 and n_1 are the refractive indices of polymer and solvent, respectively. The former was found to be 1.585 according to eq. (5).

Osmometry

Osmotic pressures Π were determined at 37° with solutions of PPSQ fractions in toluene using a Model 502 Mechrolab membrane osmometer and 08 Schleicher and Schell membranes. Only the data on fraction B-10 showed evidence of diffusion through the membrane. A Stabin osmometer was used to measure Π as a function of temperature for solutions of fractions A-4 and A-5 in ethylene dichloride. Data were plotted in the usual way according to the relation

$$\left(\frac{\Pi}{c}\right)^{1/2} = \left(\frac{RT}{M_n}\right)^{1/2}\left(1 + \frac{1}{2}\Gamma_2{}^{\Pi}c + \cdots\right) \tag{6}$$

to determine M_n and $\Gamma_2{}^{\Pi} = A_2{}^{\Pi}M_n$.

Rheometry

A cone and plate rheometer described elsewhere [9] was used to measure the steady-state viscosity η_κ as a function of the shear rate κ and the recoverable strain γ_κ^R following steady-state flow at constant shear rate. The recovery function R_κ defined by the relation

$$R_\kappa = \frac{\gamma_\kappa^R}{\kappa\eta_\kappa} \tag{7}$$

serves as a useful measure of the recovery with a limiting value R_0 at small κ equal to the steady-state recoverable compliance, often denoted by the symbol J_e^0). In a few cases the recovery $\gamma_\kappa^R(\theta)$ was measured as a function of time θ following steady-state flow. In the limit of small κ this gives the recovery $R_0(\theta)$

$$R_0(\theta) = \frac{\gamma_0^R(\theta)}{\kappa\eta_0} \tag{8}$$

as a function of θ.

Exclusion Chromatography

A Waters analytical GPC was used to obtain exclusion chromatographs with solutions of PPSQ in tetrahydrofuran. Columns with 3×10^6, 1.5×10^5, 10^4, and 10^3 Å ratings were placed in series in the order given, with the 10^3 Å column at the inlet. A 1 ml/min flow rate was used. The column set has previosuly been calibrated [10] to give the elution volume V_e in terms of the product $[\eta]M$ using fractions of anionically prepared polystyrene.

RESULTS AND DISCUSSION

Dilute Solutions

Dilute-solution parameters determined by light scattering, osmometry, and viscometry are compiled in Tables I and II. It may be seen that $[\eta]$ is only weakly dependent on the solvent, with the range of solvents including theta solvents for which A_2 is zero, and good solvents with positive A_2. For example, the data in Figure 1 show that $[\eta]$ is scarcely affected over the temperature span $45°-65°$ in ethylene dichloride, even though A_2 ranges from negative to positive over the same span [5].

TABLE I

Light-Scattering, Osmotic Pressure, and Viscosity Data for Fractions of PPSQ

Sample	$[\eta]^a$	$[\eta]^b$	$10^{-5}M_n{}^c$	$10^4A_2^{\mathrm{II}\,c}$	$10^{-5}M_w$	$10^4A_2^{\mathrm{LS}}$	$10^{12}<s^2>_{\mathrm{LS}}$
A	1.18	----	1.29	1.29	----	----	-----
A2	2.05	1.71	4.59	1.10	6.71^d	1.52^d	20.3^d
A3	1.34	1.19	2.76	1.22	2.81^e	2.14^e	6.04^e
A4	0.88	0.81	1.77	1.35	----	----	-----
A5	0.61	0.57	1.15	1.67	----	----	-----
A6	0.30	0.28	0.50	2.15	----	----	-----
B	1.04	----	0.96	----	----	----	-----
B1	2.43	1.74	4.89	1.12	9.07^d	1.45^d	27.1^d
B1	2.43	1.74	4.89	1.12	9.56^e	1.42^e	21.4^e
B2	1.87	1.50	3.64	1.14	6.31^d	1.37^d	19.2^d
B3	1.48	1.24	3.10	1.25	----	----	-----
B4	1.20	0.99	2.33	1.35	----	----	------
B5	1.03	0.86	1.94	1.46	2.57^e	2.41^e	7.42^e
B6	0.81	0.69	1.42	1.48	----	----	-----
B7	0.64	0.59	1.21	1.54	----	----	-----
B8	0.51	0.44	0.88	1.78	----	----	-----
B9	0.32_3	0.26_6	0.54	2.03	0.62^e	2.21^e	1.97^e
B10	0.12_9	0.12_5	0.26	2.43	----	----	-----

^{a} Benzene, 25°.

^{b} Ethylene dichloride, 50.5°.

^{c} Toluene, 37°.

^{d} Tetrahydrofuran, 25°.

^{e} Chloroform, 20°.

TABLE II
Dependence of Intrinsic Viscosity on Solvent

Fraction	Benzene 25°C	Chloroform 25°C	Toluene 37°C	o-Xylene 65°C	Ethylene Dichloride 50.5°C	52°C
B-1	2.43	2.33	2.21	2.11	----	----
B-2	1.87	----	----	----	1.50	1.52
B-4	1.20	1.17	1.13	1.05	0.99	1.01
B-9	0.32	0.31	0.30	0.25	0.27	0.27

The data on A_2^{II} as a function of temperature T given in Figure 1 can be analyzed to determine the thermodynamic parameter B_0 through the relations [11]

$$\lim_{T=0} A_2 \equiv A_2^L = 4\pi^{3/2}N_a B_0 \left(1 - \frac{\theta}{T}\right) \tag{9}$$

giving $B_0 = 0.69 \times 10^{-28}$ cm^{3}mol^2/g^2, which is similar in magnitude to values

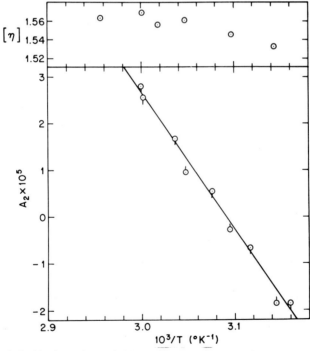

FIG. 1. Intrinsic viscosity and second virial coefficient as functions of $1/T$ for data on PPSQ in ethylene dichloride: ⊙, B-2; ⌀, A-4; and ⌀, A-5.

found with conventional flexible-chain polymers [11]—about equal, for example, with B_0 for polystyrene in 1-chloro-n-undecane. Since $[\eta]$ is nearly independent of temperature from Θ to $\Theta + 15°$, the chain dimensions change very little over the temperature range studied; that is, the expansion factor $\alpha = (\langle s^2 \rangle / \langle s^2 \rangle_0)^{1/2}$, where subscript zero denotes the unperturbed state, is nearly unity over the temperature range studied. This behavior can have either, or both, of two explanations, as follows. The expansion factor is related to an interaction parameter z by the equations

$$\alpha^2 = 1 + a_1 z + 0(z^2) \tag{10}$$

$$z = \frac{A_2^L M^{1/2}}{4\pi^{3/2} N_a (\langle s^2 \rangle_0 / M)^{3/2}} \tag{11}$$

For linear flexible chain polymers a_1 is $134/105$. One would expect a_1 to be decreased by chain stiffness, becoming zero for a truly inflexible chain [12]. Thus, intramolecular inflexibility could account for the observed closeness of α to unity, whatever the range of z. On the other hand, calculation of z for the data at 65° using the estimate $\langle s^2 \rangle_0 / M = 24 \times 10^{-18}$ (see below) gives $z = 0.02$ for polymer B-2 at 65° so that α would be expected to be near unity at 65° even if PPSQ were a flexible chain with, for example, $\langle s^2 \rangle_0$ approximately equal to the dimension for a chain with free rotation about valence bonds. For comparison, for a polystyrene chain with the same contour length and A_2^L, but with $\langle s^2 \rangle_0 / M = 7.6 \times 10^{-18}$, z would be about 0.08 so that α^2 would be about 1.1, making the effect on $[\eta]$ easily measurable.

For similar reasons, the weak dependence of $[\eta]$ on solvent shown in Table II, though consistent with the behavior expected for inflexible chains, cannot be taken as proof of rigidity. For example, over a wide range of α, the relation [11]

$$\alpha^2 - 1 = \frac{a_1 A_2 M^2}{4\pi^{3/2} N_a \langle s^2 \rangle_0^{3/2}} = f\left(\frac{A_2 M^2}{\langle s^2 \rangle^{3/2}}\right) \tag{12}$$

provides a good correlation between α and A_2. Even if a_1 has its largest value of $134/105$ or the semiempirical relation for $f(A_2 M^2 / \langle s^2 \rangle^{3/2})$ found with flexible-chain polymers [11, 13] is used, the experimental values of $A_2^{LS} M_w^2 / \langle s^2 \rangle_{LS}^{3/2}$ correspond to values of α^2 less than 1.1. Thus, the rather large value of $\langle s^2 \rangle / M$ for PPSQ is itself sufficient to reduce the effects of intramolecular interactions so that only slight chain expansions would be expected in solvents over the usual span in A_2 even if the molecule were free to undergo rotational isomerization.

Given the conclusion that chain expansion effects are small to negligible for PPSQ, some other explanations must be found for the large value of $\nu = \partial \ln [\eta] / \partial \ln M$ characterizing the data in Table I. For example, ν is about 0.92 and 0.82 for the data in benzene (25°) and ethylene dichloride (50.5°), respectively, the latter being a theta solvent for PPSQ. The nature of the dependence of $[\eta]$ on M for a wormlike chain provides a reasonable correlation and explanation of the data. According to this model, and neglecting chain expansion effects, $[\eta]$ can be expressed in the form

$$[\eta] = KL^2 f\left(\frac{L}{d_H}, \frac{\rho}{d_H}, \frac{b}{d_H}\right)$$ (13)

where $K = \pi N_a / 100 M_L$, with $[\eta]$ in dl/g. Here L is the contour length, $M_L = M/L$ is the mass per unit contour length, ρ is the persistence length, d_H is the hydrodynamic diameter, and b is the equivalent segment length. In order to compare experimental data with theoretical calculations for $[\eta]$ as a function of M, it is necessary to reduce the number of variables in eq. (13). To this end and we incorporate the Kuhn approximation in the form $M_L = m_0/l$ where m_0 and l are the molar weight and contour length of a chain repeat unit, respectively, and introduce the approximation that $b = d_H$ in models where b appears explicitly. For PPSQ, the former approximation gives $M_L = 103.2$ daltons/Å [1].

Calculations of Yamakawa and Fujii for a wormlike cylinder model [14] and of Eizner and Ptitsyn for a wormlike chain [15], with $b = d$, yield similar functions $f(L/d_H, \rho/d_H)$ over the range of parameters $L/d_H = M/M_L d_H$ and d_H/ρ of interest here. For this model

$$\frac{\langle s^2 \rangle}{M} = \frac{\rho}{3M_L} S\left(\frac{M}{M_L \rho}\right)$$ (14a)

$$S(x) = 1 - 3x^{-1} + 6x^{-2} - 6x^{-3}[1 - \exp(-x)]$$ (14b)

Comparison of eq. (13) with experimental data is facilitated by a procedure described in the Appendix. Application of that procedure should be restricted to fractions with a narrow molecular-weight distribution if possible. Here, we apply the scheme using data on $[\eta]$ versus M_{AVG} for fractions of PPSQ, with $M_{AVG} = 1.15 \times M_n \approx M_w$, since the weight-average molecular weight is most nearly appropriate when the Mark-Houwink exponent ν is nearly unity, and we have far more data on M_n than on M_w. The results, shown in Figure 2, including some data from reference [7] for comparison, give, with $M_L = 103.2$ daltons/ Å,

Benzene	Ethylene dichloride
$\rho = 74$ Å	$\rho = 73$ Å
$d_H = 8.9$ Å	$d_H = 4.4$ Å

for PPSQ. Use of these parameters in eq. (15) yields

$$\langle s^2 \rangle / M = \begin{cases} 24.0 \times 10^{-18} \text{cm}^2/\text{dalton in benzene } (25°) \\ 23.6 \times 10^{-18} \text{cm}^2/\text{dalton in ethylene dichloride } (50.5°) \end{cases}$$

These figures are in good agreement with the observed values of $\langle s^2 \rangle_{LS}/M_w$ reported here, which lie in the range 22×10^{-18} to 30×10^{-18} cm^2/dalton, and ρ is in the range usually quoted for PPSQ [4]. According to this analysis, the principal difference between values of $[\eta]$ in the "good solvent" benzene and the theta solvent ethylene dichloride (50.5°) is in the value of d_H. The invariance of ρ for PPSQ in solutions for which A_2 is so different is in accord with the premise that PPSQ is an inflexible-chain polymer but can also be explained on the basis of the small values of z for PPSQ, as discussed above.

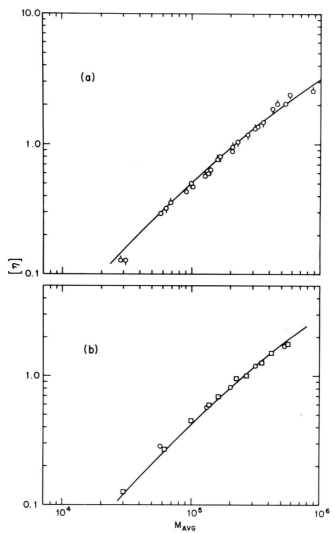

FIG. 2. Intrinsic viscosity versus average molecular weight $M_{AVG} = 1.15\,M_n \simeq M_w$ for fractions of PPSQ in (a) benzene (25°) and (b) ethylene dichloride (50.5°): O, series A; □, Q, series B; and Ò, data from reference 7. The curves represent theoretical calculations corresponding to the values of ρ and d_H given in the text.

The data on A_2^{II} as a function of M_n show that A_2^{II} decreases with increasing M_n, with $\gamma = \partial \ln A_2^{II}/\partial \ln M_n$ equal to -0.27 for PPSQ in toluene. This differs from a report that γ is about equal to zero for solutions of PPSQ in benzene [7]. Yamakawa and Stockmayer [12] have published a perturbation theory for A_2 for a wormlike chain that leads to the result

$$A_2 = A_2^L h(z) \tag{15a}$$

$$h(z) = 1 - Q\left(\frac{L}{\rho}, \frac{d_T}{\rho}\right) z + \cdots \tag{15b}$$

where d_T is the effective thermodynamic diameter of the chain. If the model is strictly applied and a hard-sphere potential between chain segments is assumed, then

$$A_2^l = \frac{2\pi N_A d_T}{3M_L^2} \tag{16}$$

Since the series expansion in eq. (15) is not rapidly convergent, we will use the approximation

$$h(z) \cong \frac{1 - \exp(-2Qz)}{2Qz} \tag{17}$$

to evaluate $h(z)$ [11]. Contributions to intramolecular chain expansion are not included in eq. (17) on the basis that they are unimportant either because PPSQ is an inflexible-chain polymer or because z is very small. When Qz is greater than about 1.5, $h(z)$ is approximately $(2Qz)^{-1}$ according to eq. (17), so that $A_2 M^{1/2}$ tends to proportionality with Q^{-1}:

$$A_2 M^{1/2} = 4\pi^{3/2} N_a \left(\frac{\langle s^2 \rangle_0}{M}\right)^{3/2} \frac{1 - \exp(-2Qz)}{2Q} \tag{18}$$

Thus, for $Qz > 1.5$, it is not possible to evaluate A_2^l from A_2 without the use of some model, for example, by the use of eq. (16). The data in Figure 3 show that reasonable agreement between this approximate theory and the data on PPSQ in toluene is achieved with $d_T/2\rho = 0.009$, or $d_T = 1.4$ Å with $\rho = 75$ Å. With this value of d_T, eq. (16) gives $A_2^l = 1.66 \times 10^{-4}$, so that z is found to be less than 0.1 for solutions of PPSQ in toluene for all the fractions used here, confirming the judgment made in the preceding analysis that the z is small enough to account for negligible to small chain expansion effects, whatever the chain flexibility.

With several of the polymers M_w was determined from the exclusion chromatography data using the correlation between $[\eta]M$ and V established with polystyrene solutions [10] and the correlation between $[\eta]$ and M given in Figure 1(a). The apparent distribution curves were not corrected for axial diffusion, and the usual relations were used to calculate M_w, e.g., $M_w = \Sigma H_i M_i / \Sigma H_i$, with H_i the observed chart reading for species i, etc. The results for M_w were in satisfactory agreement with M_w determined by light scattering, e.g., $10^{-5} M_w$ equal to 5.9, 10.0, and 6.3 for samples A-2, B-1, and B-2, respectively. Thus, it appears that the "universal correlation" between $[\eta]M$ and V_e useful with many flexible-chain polymers is also useful with PPSQ.

We conclude that on the basis of the dilute solution properties studied here PPSQ can be modeled as a wormlike chain with a persistence length of about 75 Å, or with a Kuhn segment with length 150 Å. This is in accord with conclusions reached by others on the basis of similar data [4, 7, 16]. It should be noted that our results do not indicate whether PPSQ is an inflexible chain or has some backbone bonds that permit rotational isomerization, e.g., incompletely cyclized repeat units. It would appear, however, that there is no need to postulate

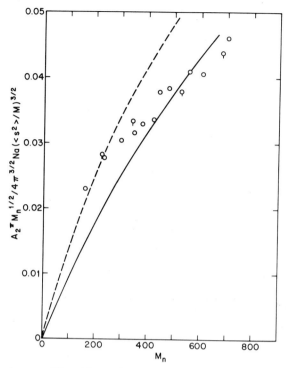

FIG. 3. The function $A_2 M_n^{1/2}/4 \Pi^{3/2} N_a (\langle s \rangle / M)^{3/2}$ versus $M_n^{1/2}$ for fractions of PPSQ in toluene: O, series A, and Q, series B. The solid and dashed lines are calculated with eq. (18) for $d_T/2\rho$ equal to 0.009 and 0.012, respectively.

the existence of branched-chain configurations to explain the light-scattering and viscometric data on the polymers studied here. Branched configurations have been cited previously to explain anomalous viscosity–molecular-weight behavior [17–18] and gelation phenomena [17], but it seems unlikely that the consistent behavior observed here for data on fractions of two separate polymers would have been obtained if these polymers had any reasonable amount of branching. In addition, solubility was not a problem with the polymers studied here. Consequently, we can conclude that the PPSQ studied here has substantially the ladder structure depicted above, recognizing, however, that the data reported on dilute solutions would not be sensitive to a small extent of incompletely cyclized repeating units.

Concentrated Solutions

Plots of η_κ and R_κ versus κ for solutions of PPSQ fractions in cyclohexanone are given in Figure 4. In studies with flexible-chain polymers, it has been found that η_κ and R_κ can conveniently be represented in the forms [19, 20]

$$\eta_\kappa = \eta_0 Q(\tau_c \kappa) \tag{19}$$

$$R_\kappa = R_0 P(\tau_c \kappa) \tag{20}$$

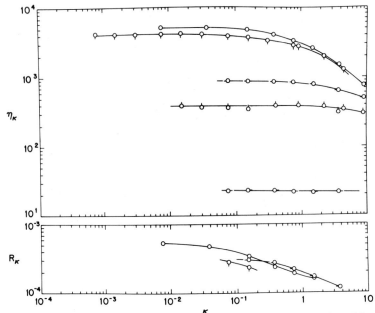

FIG. 4. Viscosity and recovery functions versus rate of strain for solutions of PPSQ in cyclohexanone (0.20 g/ml): φ, A-2 (15°); O, A-2 (27°); ⊖, A-3 (8.5°); ⊖, A-3 (27°); and O-, A-5 (29°).

where $\tau_c = R_0\eta_0$ is a characteristic time. Equation (19) has the form expected for a thermorheologically simple fluid [21]. The data obtained here with PPSQ solutions are plotted in reduced form in Figure 5 and compared with similar data observed with a 13% polystyrene solution [20]. The correspondence observed between reduced flow curves (η_κ/η_0 versus $\tau_c\kappa$) for solutions of PPSQ and polystyrene does not imply that the two polymers have similar flexibility but only that the time constant $R_0\eta_0$ is a useful measure of the time scale of molecular motion in a shear gradient.

For flexible-chain polymers, η_0 can be correlated by the relation [22]

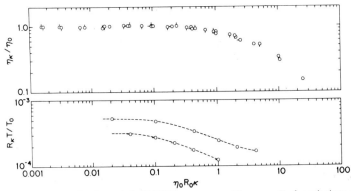

FIG. 5. Reduced plots of η_κ/η_0 and $R_\kappa(T/288)$ versus $\tau_c\kappa$ with $\tau_c = \eta_0 R_0$ for solutions of PPSQ identified in the caption to Figure 4. The dashed lines show data for $R_0(\theta)$ versus τ_c/θ for fractions A-2 and A-3.

$$\eta_0 = \left(\frac{N_a}{6}\right) X \left(\frac{X}{X_c}\right)^a \zeta \tag{21}$$

where

$$X = \frac{\langle s^2 \rangle_0 c}{m_a} \tag{22}$$

$$a = \begin{cases} 0 & \text{if } X \leq X_c \\ 2.4 & \text{if } X > X_c \end{cases}$$

Here, m_a is the molar weight per hydrodynamic unit with friction factor ζ. For nearly all systems studied so far, the temperature dependence of ζ can be represented with the Vogel relation

$$\zeta = \zeta_0 \exp\left[\frac{C}{(T - T_0)}\right] \tag{23}$$

For many flexible-chain polymers, X_c is about 400×10^{-17} if the number of hydrodynamic units is equated with the number of backbone chain atoms [22]. According to eq. (22), insofar as $\langle s^2 \rangle_0 / M$ is a constant, data on η_0 versus cM should be on a single curve if ζ is held constant, with $\eta_0 \propto (cM)^{3.4}$ for $cM > \rho_2 M_c$ and $\eta_0 \propto cM$ for smaller cM, where ρ_2 is the density of the bulk polymer and M_c is a constant equal to $X_c m_a / (\langle s^2 \rangle_0 / M) \rho_2$. Data for solutions of PPSQ fractions in benzene at 30°, with concentrations ranging from 0.05 to 0.1 g/ml are plotted in this way in Figure 6. It is seen that the familiar 3.4 and 1.0 power laws are observed with the PPSQ solutions, with $\rho_2 M_c$ equal to 9120. Since all the data lie on one curve, it appears that ζ is independent of concentration for PPSQ in benzene over the concentration range studied. This is not unusual at concentrations up to 0.1 g/ml if $T > T_0$, since C is usually only a weak function of c, whereas T_0 may depend markedly on c [22].

Data on solutions of PPSQ in cyclohexanone with $c = 0.2$ g/ml also shown in Figure 6 exhibit the 3.4 power law dependence but do not extend to low enough cM to reach the critical point $cM = \rho_2 M_c$. Since the apparent activation energy $E_a = R \partial \ln \eta_0 / \partial (T^{-1})$ is concentration-dependent in this range, we could not study η_0 at fixed ζ by simply using less concentrated solutions, e.g., E_a is 5.45 and 3.80 kcal/mole for 0.2 and 0.1 g/ml solutions in cyclohexanone, respectively.

With the value of $\langle s^2 \rangle / M$ given above we find $m_a X_c = 2.30 \times 10^{-13}$. The significance of m_a is not clear for a ladder polymer like PPSQ. Whereas ζm_a^{-1} is uniquely defined, separation into the terms ζ and m_a requires the assignment of a hydrodynamic unit. With flexible-chain polymers it has been found that X_c is about 400×10^{-17} for a wide variety of polymers if m_a is defined by the number of backbone chain atoms, irrespective of the size or nature of side chain substituents. If a similar procedure is followed here, with each "rail" of the ladder presumed to contribute to ζ, then $m_a = m_0/4$, where $m_0 = 258$ is the molar weight per repeat unit, and $X_c = 356 \times 10^{-17}$ for PPSQ in benzene. Thus, with respect to both the general form of the dependence of η_0 on cM and the value of X_c, it appears that at least in moderately concentrated solutions PPSQ behaves very much like flexible coil chains.

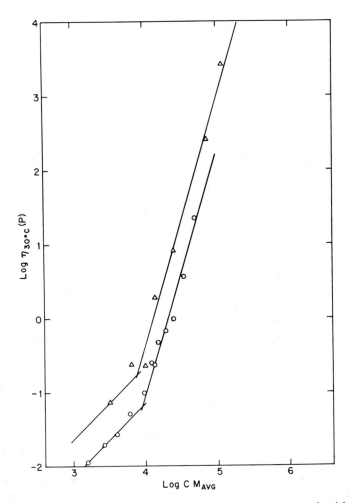

FIG. 6. Viscosity η_0 (at 30°) versus cM_{AVG} for solutions of PPSQ in benzene, O, and cyclohexanone, $\triangle$. M_{AVG} is equal either to 1.15 $M_n \approx M_w$ or to M_w.

It is of interest to compare the characteristic time $\tau_c = \eta_0 R_0$ for PPSQ with the value that would be expected for a flexible-chain polymer and with the rotatory relaxation time $\tau_R = (2D_R)^{-1}$ for a rigid chain, where D_R is the rotatory diffusion constant given by $\eta D_R = KRT/M[\eta]$, with K a constant dependent on the shape of the rigid molecule (particle) [23]. For monodisperse flexible-chain polymers, it appears that R_0 can be expressed in the form [19, 24]

$$c^2 (R_0)_{\text{flex}} = \frac{2}{5} \frac{E^* \rho_2 M_2}{RT} \tag{24}$$

if the entanglement density $E = cM/\rho_2 M_e$ exceeds a critical value E^* of about 10. Here, M_e is the molecular weight between entanglement loci ($M_e \cong M_c/2$). Putting $E^* M_e = E^* M_c/2 = 5M_c$, we find that $(R_0)_{\text{flex}} = 0.2 \times 10^{-4}$ cm^2/dyn with eq. (24) for a 0.2 g/ml solution of PPSQ, compared with the value of

$3 - 5 \times 10^{-4}$ cm^2/dyn observed for samples A-2 and A-3. Thus, τ_c observed with PPSQ is tenfold larger than the value $\eta_0(R_0)_{flex}$ one would normally expect for a monodisperse linear flexible chain. This seems reasonable in view of the presumed inflexibility of PPSQ, but it should be remarked that a long high-molecular-weight tail would have a similar effect on τ_c for a flexible-chain polymer.

On the other hand, for a rigid coil, for which K is $1/4$, compared with $2/15$ for a rigid rod, $\tau_R/\eta = 2M[\eta]/RT$ is equal to 115×10^{-4} and 30×10^{-4} cm^2/dyn for fractions A-2 and A-3, respectively. Electric birefringence studies on dilute solutions of PPSQ show that values of τ_R calculated in this way agree satisfactorily with those determined from dispersion curves, indicating that PPSQ behaves as a rigid coil in such experiments [23]. Values of τ_R/η so calculated are much greater and more dependent on M than the observed τ_c/η_0. By comparison, for a dilute suspension of rigid rods, theoretical calculations lead to the result $\tau_c = (3/5)\tau_R$ [25, 26]. It appears, therefore, that τ_c observed with concentrated solutions of PPSQ is larger and more dependent on M than would be expected for a flexible-chain polymer with $cM > k\rho_2 M_c$ but is smaller and less dependent on M than would be expected for noninteracting rigid coils. It is possible that τ_c is proportional to an averaged τ_R, the latter reflecting the presence of low-molecular-weight components in the fraction, since these have the greatest mobility. On the other hand, it may be that τ_c, and R_0, reflect the flexing of bonds in the PPSQ backbone or the presence of imperfectly cyclized repeat units.

In addition to the recovery function R_x, we have obtained limited data on the transient recovery $R_0(\theta)$ as a function of the recovery time θ following steady-state creep at a stress low enough to ensure a linear response. In previous studies with flexible-chain polymers we have observed that the function R_0 ($\theta = x^{-1}$) bears a close resemblance to x for θ in the range $\theta > \tau_c$. A similar correspondence is noted here, as may be seen in Figure 5, where the dashed line gives $R_0(\theta)$ versus τ_c/θ for comparison with R_x as a function of $\tau_c x$. With monodisperse flexible-chain polymers, $R_0(\theta)$ is often equal to its long-time limit R_0 if $\theta/\tau_c > 1$, or R_x is equal to R_0 if $\tau_c x < 1$ [19]. With these systems, molecular-weight heterogeneity causes more pronounced dependence on θ, or x. The behavior observed with PPSQ is not appreciably different from that for fractionated flexible-chain polymers over the limited time scale studied.

CONCLUSIONS

Studies on dilute solutions of PPSQ show that data on A_2, $[\eta]$, and $\langle s^2 \rangle$ as functions of molecular weight, temperature, and solvent can be understood in terms of a wormlike chain model with a persistence length ρ of about 75 Å. Intramolecular chain expansion effects are negligible, owing either to a large unperturbed chain dimension or to inherent chain backbone rigidity—the data on A_2, $[\eta]$, and $\langle s^2 \rangle$ cannot distinguish between these possibilities. Values of the single-contact term A_2^t in expressions for A_2 are similar to those observed with the more widely studied flexible-chain polymers.

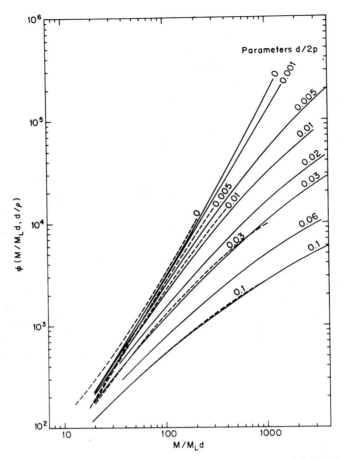

FIG. 7. The functions φ_{YF}, —, and φ_{EP}, - - - -, calculated by Yamakawa and Fujii [14] and Eizner and Ptitsiyn [15], respectively, for the indicated values of ρ.

The rheological data show that the solutions of PPSQ studied have properties qualitatively very similar to the properties of flexible-chain polymers with one exception. Thus, the familiar dependence of η_0 on $(cM)^{3.4}$ or cM for cM greater or less than $\rho_2 M_c$, respectively, is observed, and the reduced-flow curve η_κ/η_0 versus $\tau_c \kappa$ for the PPSQ solutions studied is similar to reduced-flow curves for flexible-chain polymers. It is found, however, that the characteristic time $\tau_c = \eta_0 R_0$ is longer and more dependent on M for the PPSQ solutions than would be expected for a flexible-chain polymer. The longer correlation time observed for PPSQ may reflect the inflexibility of the PPSQ molecule.

It is a pleasure to acknowledge the technical assistance of T. A. Knell in some of the measurements. T. E. Helminiak was a Visiting Fellow at Mellon Institute, now part of Carnegie-Mellon University, during this study. The study was supported in part by grant CP-28538X from the National Science Foundation.

APPENDIX

For the models of interest here, eq. (13) of the text can be written in the form

$$[\eta] = \frac{\pi N_A (M_L d)^2}{100 M_L^3} [\eta]_R \tag{25a}$$

$$[\eta]_R = K' \phi \left(\frac{M}{M_L d}, \frac{d}{2\rho} \right) \tag{25b}$$

where K' is a numerical constant and $\phi(M/M_L d, d/2\rho)$ is a function of the indicated variables. The functions ϕ_{YF} and ϕ_{EP} calculated by Yamakawa and Fujii [14] and Eizner and Ptitsyn [15], respectively, are displayed in Figure 7. For the former, the values of ϕ_{YF} given numerically in reference were augmented by interpolation of ϕ versus $d/2\rho$ at constant $M/M_L d$ for the curves with $d/2\rho$ equal to 0.02 and 0.06. The functions ϕ_{YF} and ϕ_{EP}, calculated with K' equal to $1/24 = 0.0417$ and 0.0473, respectively, are in good agreement. In analysis of experimental data on $[\eta]$ versus M, a graph of $100 M_L^3 [\eta]/\pi N_A K'$ versus M may be superposed on Figure 7 and translated along a line of slope 2 (the dashed line) until agreement between the data and one of the theoretical curves at constant $d/2\rho$ is achieved. The value M^* of M for which $M/M_L d$ is unity then gives $d = M^*/M_L$, and ρ may be calculated from the value of $d/2\rho$ and d. The parameters given in the text were determined with $K' = 1/24$.

REFERENCES

[1] J. F. Brown, Jr., L. H. Voght, Jr., A. Katchman, J. W. Eustace, K. M. Kiser, and K. W. Krantz, *J. Am. Chem. Soc., 82,* 6194 (1960).

[2] V. N. Tsvetkov, *Russ. Chem. Rev., 38,* 755 (1969).

[3] V. N. Tsvetkov, *Eur. Polym. J. Suppl.* 237, (1969).

[4] V. N. Tsvetkov, *Polym. Sci. USSR, 16,* 1087 (1974).

[5] T. E. Helminiak, C. L. Benner, and W. E. Gibbs, *ACS Polym. Prepr., 8*(1), 284 (1967).

[6] E. F. Casassa and G. C. Berry in "Polymer Molecular Weights," Part 1, P. E. Slade, Jr., Ed., Dekker, New York, 1975, Chap. 5.

[7] V. Ye. Eskin, O. Z. Korotkina, P. N. Lavrenko, and Ye. V. Korneyeva, *Polym. Sci. USSR, 15,* 2391 (1973).

[8] V. N. Tsvetkov, K. A. Andrianov, G. I. Okhrimenko, and M. G. Vitovskaya, *Eur. Polym. J., 7,* 1215 (1971).

[9] G. C. Berry and C.-P. Wong, *J. Polym. Sci. Polym. Phys. Ed., 13,* 1761 (1975).

[10] G. C. Perry, *J. Polym. Sci. Part A-2, 9,* 687 (1971).

[11] G. C. Berry and E. F. Casassa, *J. Polym. Sci. Part D, 4,* 1 (1970).

[12] H. Yamakawa and W. H. Stockmayer, *J. Chem. Phys., 57,* 2843 (1972).

[13] D. W. Tanner and G. C. Berry, *J. Polym. Sci. Polym. Phys. Ed., 12,* 941 (1974).

[14] H. Yamakawa and M. Fujii, *Macromolecules, 7,* 128 (1974).

[15] Y. E. Eizner and O. B. Ptitsyn, *Vysokomol. Soedin., 4,* 1725 (1962).

[16] W. H. Stockmayer, *Br. Polym. J., 9,* 89 (1977).

[17] C. L. Frye and J. M. Klosowski, *J. Am. Chem. Soc., 93,* 4599 (1971).

[18] J. Kouar, L. Mrkvicková-Vaculová, and M. Bohdanecký, *Makromol. Chem., 176,* 1829 (1975).

[19] G. C. Berry, B. L. Hager, and C.-P. Wong, *Macromolecules, 10,* 361 (1977).

[20] C.-P. Wong and G. C. Berry, *Polym. Prepr. Am. Chem. Soc. Div. Polym. Chem., 17*(2), 413 (1976); *15*(2), 126 (1974).

[21] H. Markovitz, *J. Polym. Sci. Polym. Symp., 50,* 431 (1975).

[22] G. C. Berry and T. G Fox, *Adv. Polym. Sci., 5,* 261 (1968).

[23] V. N. Tsvetkov, K. A. Andrianov, E. I. Riumtsev, I. N. Shtennikova, N. V. Pogodina, G. F. Kolbina, and N. N. Makarova, *Eur. Polym. J., 11,* 771 (1975).

[24] W. W. Graessley, *Adv. Polym. Sci., 16,* 1 (1974).

[25] J. G. Kirkwood and J. Plock, *J. Chem. Phys., 24,* 665 (1956).

[26] T. Kotaka, *J. Chem. Phys., 30,* 1566 (1959).

PROPERTIES OF A PHENYL-SUBSTITUTED POLYPHENYLENE IN DILUTE SOLUTION

JAMES L. WORK,* GUY C. BERRY, and
EDWARD F. CASASSA†
Department of Chemistry, Carnegie-Mellon University, Pittsburgh, Pennsylvania 15213

JOHN K. STILLE‡
Department of Chemistry, The University of Iowa, Iowa City, Iowa 52240

SYNOPSIS

Three preparations of a phenyl-substituted polyphenylene were fractionated and studied in solution by light scattering, osmometry, viscometry, fluorescence depolarization, and exclusion chromatography. Three of the five main-chain phenylene groups in the repeat unit are linked para, but the other two may be either meta or para, with the latter linkage predominating. The data are consistent with a chain structure characterized by rigid segments about 70–75Å long, on the average, joined with 120° bond angles, about which virtually free rotation is allowed. For chains of up to about 250 backbone phenylene units, the hydrodynamic behavior in tetrahydrofuran, and other solvents, is that of a freely draining coil with intrinsic viscosity proportional to molecular weight. The effect of intramolecular excluded volume on chain conformation is negligible, although intermolecular interactions are reflected by a positive second virial coefficient (in tetrahydrofuran). Study of polymers of higher molecular weight is hampered by formation of what appear to be metastable aggregates resembling randomly branched chains.

INTRODUCTION

Certain polymers synthesized in recent years are characterized by long, relatively bulky, inflexible backbone elements. Though these units may be linked together by single bonds that allow sufficiently long chains to assume the gross conformation of a flexible coil, the polymers typically exhibit limited solubility, are soluble only in protonated form in strong acids, and are intensely colored.

---

* Present address: Armstrong Cork Company, Lancaster, Pennsylvania 17604.

† To whom inquiries may be addressed.

‡ Present address: Department of Chemistry, Colorado State University, Fort Collins, Colorado 80523.

Journal of Polymer Science: Polymer Symposium 65, 125–141 (1978)
© 1978 John Wiley & Sons, Inc.

0360-8905/78/0065-0125$01.00

These properties make detailed study of dilute solutions difficult or impossible and frustrate comparison with theory for the behavior of the more familiar flexible chains.

We, therefore, considered it of interest to study a substituted polyphenylene that gives nearly colorless solutions in a variety of ordinary organic solvents [1] and yet has long rodlike segments in the chain backbone. The structure incorporate five backbone phenylene rings per repeat unit:

Three of the five are linked para, but the second and fourth, as shown above, can be either para or meta. Studies on formation of model compounds suggest that the alternative catenations are about equally probable [2]. Lack of steric hindrance at the link between repeat units renders the chain flexible in the sense that rotational isomerization is possible for any chain with more than one meta linkage.

In the following we report dilute solution measurements on fractionated polymer and explore correlations among the results from exclusion chromatography, osmotic pressure, light scattering, intrinsic viscosity, and fluorescence depolarization.

EXPERIMENTAL

Materials

A Diels-Alder step-growth reaction was used to obtain the polymer. The synthesis, from the preparation and purification of the monomer, is described elsewhere [2, 3]. The polymer was fractionated by precipitation from chloroform solution (1.0 g/liter) by addition of methanol. The nonsolvent was added slowly to bring the stirred solution to the cloud point. The solution was heated until it was again clear; a small additional amount of solvent was added; and then the solution was cooled slowly to the original temperature, whereupon agitation was stopped and the precipitated polymer was allowed to settle. In each case, some 12 primary fractions collected by repetition of this procedure were refractionated to provide 3 or 4 sharper secondary fractions. These were dissolved in benzene and freeze-dried.

Three polymer preparations, designated 10-71, 9-68, and 8-73 are discussed in the following. They differ in molecular weight, sample 10-71 being the lowest polymer. The fraction designation is added to the polymer code: 10-71; 4B, for example, denoting the second secondary fraction (B) of the fourth primary fraction of the polymer 10-71.

Reagent-grade tetrahydrofuran (THF), containing about 0.025% inhibitor

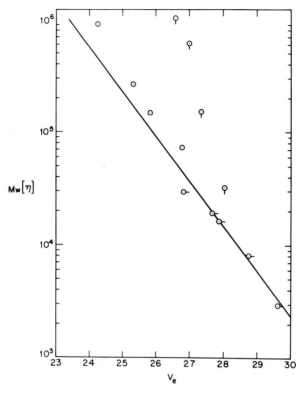

FIG. 1. Dependence of $M_w[\eta]$ on elution volume for fractions from poly-phenylene samples 9-68 (⊙), 10-71 (⊙-), and 8-73 (♀). The solid line is the calibration curve established with narrow distribution polystyrene.

(butylated hydroxytoluene), was used for exclusion chromatography, osmom-etry, viscometry, and light scattering. Reagent-grade toluene and benzene and redistilled quinoline and cyclohexanone were also used for viscometry. Re-agent-grade dimethyl phthalate was the solvent for fluorescence depolarization measurements.

Exclusion chromatography

A Waters Associates analytical gel-permeation chromatograph was used to characterize the fractions. Four Styragel columns with nominal pore-size ratings of 10^3, 10^4, 1.5×10^5, and 3×10^6 were put in series in this order. The elution volume V_e corresponding to the maximum in the refractive index of the effluent was determined from the recorded output of the refractometer incorporated in the instrument. The column set was calibrated by correlating the product of molecular weight M and intrinsic viscosity $[\eta]$ with elution volume using plots of log $\{M[\eta]\}$ versus V_e for linear narrow-distribution polystyrene [4]. The calibration so obtained [5] is shown in Figure 1.

Elution was carried out with tetrahydrofuran at a flow rate of 1 ml/min. In every run, 0.5 ml of a 4 g/liter solution was injected. Samples were run in duplicate and gave elution volumes reproducible to 0.12 ml or better. A polystyrene sample was injected at intervals to confirm that column characteristics had not changed with time.

Osmometry

Number-average molecular weights M_n were determined from measurements at 25° with a Mechrolab Model 503 osmometer equipped with a Schleicher and Schuell Type 0-7 membrane. Solutions in THF for these and other measurements were made up by weighing. Weight-volume concentrations c were calculated on the assumption of no volume change on mixing. Concentrations were chosen to give osmotic pressures Π no greater than 3 cm of THF. No diffusion of polymer through the membrane was encountered. Data were analyzed graphically by plotting $(\Pi/c)^{1/2}$ versus c [6]. The dependence was found to be linear, and the plots were extrapolated to $c = 0$ to obtain the molecular weight from the limiting relation $(\Pi/c)_{c=0} = RT/M_n$, in which RT has the usual significance.

Light Scattering

The light-scattering photometer has been described elsewhere [7]. In all measurements, vertically polarized 5461-Å incident light was used with a band-pass interference filter placed before the detector photomultiplier to eliminate any fluorescence scattering. There was no measurable depolarization of scattered light and negligible absorption with the solutions used. A thermostat maintained the sample temperature at 25° within ±0.05°. Solutions were clarified by filtration through a Gelman Polypore membrane with a 0.4-μm pore-size rating.

The weight-average molecular weight M_w and the light-scattering averages of the second virial coefficient A_2^{LS} and the mean-square molecular radius of gyration $\langle s^2 \rangle_{LS}$ were obtained by analysis of data according to the usual relations [6,8,9]:

$$\lim_{\theta=0} \left[\frac{Kc}{R(\theta,c)} \right] \equiv \frac{Kc}{R(0,c)} = \frac{1}{M_w} + 2A_2^{LS}c + \cdots \tag{1}$$

$$\lim_{c=0} \left[\frac{Kc}{R(\theta,c)} \right] \equiv \frac{1}{M_w} \left[1 + \frac{1}{3} \left(\frac{4\pi n}{\lambda} \right)^2 \langle s^2 \rangle_{LS} \sin^2 \frac{\theta}{2} + \cdots \right\} \tag{2}$$

where K is a constant of the system at a given wavelength and $R(\theta,c)$ is the reduced intensity at concentration c and scattering angle θ. Plots of $Kc/R(\theta,c)$ versus $\sin^2 (\theta/2)$ at fixed concentration were extrapolated to $\theta = 0$ to obtain $Kc/R(0,c)$, and these values were plotted against c to obtain M_w and A_2^{LS} from eq. (1). Similarly, plots of $Kc/R(\theta,c)$ versus c at fixed θ were extrapolated to c

= 0, and the intercepts were plotted against $\sin^2 (\theta/2)$ to obtain $\langle s^2 \rangle_{LS}$. Four concentrations were used in each series. The concentrations were checked by evaporation to dryness.

The refractive index increment dn/dc is needed to calculate K. It was determined for 5461-Å light by using a differential refractometer [7,9]. Values of 0.294 and 0.250 ml/g were found for solutions in THF and benzene, respectively, at 25°.

Viscometry

Suspended-level Ubbelohde viscometers (Cannon Instrument Co.) were used for all measurements with the exception noted below. The viscometers were immersed in a 25° constant temperature bath (±0.02°). Data on four concentrations of each sample were plotted against c according to the relations [6]

$$\eta_{sp}/c = [\eta] + k'[\eta]^2 c + \cdots \tag{3}$$

$$(\ln \eta_{rel})/c = [\eta] - \left(\frac{1}{2} - k'\right) [\eta]^2 c + \cdots \tag{4}$$

to obtain the intrinsic viscosity $[\eta]$ from the intercepts and k' from the slopes. Here, η_{rel} and η_{sp} are, respectively, t/t_0 and $(t - t_0)/t_0$, with t the flow time of solution and t_0 that of solvent. Flow times were corrected for kinetic energy effects, as necessary. Values of the Huggins constant k' were between 0.4 and 0.5.

Measurements on several polymer fractions with a low-shear viscometer [10] revealed no appreciable effect of rate of shear.

Fluorescence

Absorption and fluorescence spectra of THF solutions of the substituted polyphenylene were obtained, respectively, with a Cary Model-14 spectrophotometer and an Aminco-Bowman fluorometer.

Fluorescence depolarization of solutions of the polyphenylene (1.7 g/liter) in dimethylphthalate was determined with the light-scattering photometer, using vertically polarized 4358-Å incident light and a 4358-Å band-reject interference filter before the detector. An analyzer before the photomultiplier permitted measurement of the vertical ($I_\parallel$) and horizontal (I_+) fluorescence intensity components excited by the vertically polarized incident light. Since data analysis required only the dimensionless fluorescence emission anisotropy [11]

$$r = (I_\parallel - I_+)/(I_\parallel + 2I_+) \tag{5}$$

it was not necessary to have absolute intensities. The fluorescence anistropy was measured over the temperature interval 25°–150°. The temperature dependence of the solvent viscosity, which is required for analysis of the fluorescence data, was measured with an Ubbelohde viscometer.

WORK ET AL.

TABLE I
Dilute Solution Parameters for Phenyl-Substituted Polyphenylene in
Tetrahydrofuran at 25°

Polymer	$10^{-4}M_w$	$10^{-4}M_n$	$10^{12}\langle s^2 \rangle^{LS}$ (cm^2)	$10^4 A_2^{LS}$ (dl/g^2)	$[\eta]$ (dl/g)
10-71; 4A	3.92	2.37	3.52	2.1	0.785
3	2.96	2.19	3.41	5.4	0.665
4B	2.73	2.15	2.63	4.0	0.601
5B	2.00	1.52	3.39	9.9	0.410
7B	1.13	0.96	1.41	11.8	0.266
8B	—	—	—	—	0.215
9B	—	—	—	—	0.170

DISCUSSION

Molecular Weights and Exclusion Chromatography

Dilute solution characterization results for fractions of polymer 10-71 in THF are collected in Table I.

From Figure 1, it can be seen that exclusion chromatography of fractions of this polymer follows the "universal" calibration [4]: the elution peak volumes V_e correlate with the product $M_w[\eta]$ just as they do for polystyrene. However, fractions from polymers 7-68 and 8-73 deviate systematically from that correlation and from each other. Deviations of this sort suggest an association that occurs at concentrations used in the determination of M_w by light scattering but is absent, or decreased, at the much lower concentration at which polymer is eluted from the chromatographic column. Unless association produces aggregates comparatively stable toward dilution, it might be detected in light-scattering experiments by an apparent dependence of chain dimensions on concentration. Such behavior was sometimes discerned with solutions of polymers 9-68 and 8-73. In some instances, the apparent molecular weight depended on the procedure used in making up the solution, a sign that metastable aggregates had formed.

The chromatogram of unfractionated polymer 10-71 was symmetrical; and the ratio M_w/M_n estimated from the dispersion, ignoring axial diffusion, was 3.9. Polymers 9-68 and 8-73 are apparently of higher molecular weight; the chromatogram of 9-68 was skewed toward high molecular weight and that of 8-73 was bimodal. In view of the presumptive association complicating the behavior of these two polymers, we have concentrated on analysis of data from fractions of polymer 10-71, which appears to be free of this complication.

Molecular Dimensions

The mean-square molecular radius of gyration of the polyphenylene necessarily depends on the proportions of meta and para catenation. The light-scattering data in Table I yield an estimate

$$\langle s^2 \rangle_{LS}/N_z a^2 = 8.5 \pm 1.7 \tag{6}$$

for the characteristic ratio of $\langle s^2 \rangle$ to the root-mean-square end-to-end distance of a hypothetical freely jointed chain of N_z links, each of length a. Here, N_z is taken as the z-average number of backbone phenylenes and hence $a = 4.4$ Å is the length of a phenylene unit along the chain axis. The value of N_z was estimated from M_n and M_w on the assumption that the Schulz-Zimm distribution function [8] applies, i.e., that $N_z = M_z/167 = M_w(h + 2)/167 (h + 1)$, where $h = M_w/M_n$. The ratio $\langle s^2 \rangle_{LS}/N_z$ is sensibly equivalent to $\langle s^2 \rangle/N$ for the homogeneous polymer with $N = N_z$ [9].

Equation (6) is to be compared with what would be expected for all-para catenation (a rigid rod)

$$\langle s^2 \rangle/Na^2 = N/12 \tag{7}$$

and for meta catenation of the two disposable bonds in each repeat unit. In the latter case, the chain is a sequence of links alternating in length, $l = 2a$ and $m = 3a$, joined by a valence angle $\pi - \alpha$ of $2\pi/3$ radians, which gives, for large N [12],

$$\langle s^2 \rangle = \frac{2N}{5} \left(\frac{l^2 + m^2}{12} \right) \left[\frac{1 + \cos \alpha}{1 - \cos \alpha} - \frac{2(l - m)^2}{l^2 + m^2} (\cos \alpha + \cos^3 \alpha) \right] \tag{8}$$

$$\langle s^2 \rangle/Na^2 = 1.258 \tag{9}$$

where N is still the number of phenylene units in the chain. Consequently, we must explain a characteristic size parameter that appears to be independent of chain length, as with the all-meta model, but has a value much larger than the predicted value. Granting free rotation about the axis of the unsubstituted phenylene unit, we approach the problem by assuming that the experimental $\langle s^2 \rangle/Na^2$ results from a random distribution of para and meta linkages occurring with respective probabilities p and $1-p$ at the disposable sites.

Two variants of this king of model are of interest. First we represent the chain by $n = 5N/2$ equivalent bonds of mean-square length $\langle l^2 \rangle = \frac{1}{2}[(2a)^2 + (3a)^2]$ connected by valence angles $\pi - \alpha$, where α is zero or $\pi/3$ with probabilities p and $1-p$. For large n we then have

$$\langle s^2 \rangle = n \langle l^2 \rangle \left(\frac{1 + \langle \cos \alpha \rangle}{1 - \langle \cos \alpha \rangle} \right) \tag{10}$$

and thus

$$\frac{\langle s^2 \rangle}{Na^2} = \frac{13}{30} \left(\frac{1 + \langle \cos \alpha \rangle}{1 - \langle \cos \alpha \rangle} \right) \tag{11}$$

with

$$\langle \cos \alpha \rangle = (1 + p)/2 \tag{12}$$

For the second, and more realistic, calculation we let bonds of mean-square length $\langle l^2 \rangle$ comprising segments with all-para catenation be connected by a valence angle $\pi - \alpha = 2\pi/3$. Then, the assumption of randomness in the distribution of meta and para linkages permits calculation of $\langle l^2 \rangle$. For example, the probability that a straight sequence is terminated at length $2a$ or $3a$ is $1 - p$; the probability of termination at $4a$ is zero; the probability of termination at $5a$ is $2p(1 - p)$; etc. In this way the asymptotic relation for large n

$$\frac{\langle l^2 \rangle}{a^2} = \frac{(1 - p) \left[(2^2 + 3^2) + 2(5^2)p + (7^2 + 8^2)p^2 + 2(10^2)p^3 + \cdots \right]}{2(1 - p) \left[1 + p + p^2 + p^3 + \cdots \right]}$$

$$= \frac{1}{2}(1 - p) \sum_{n=0}^{\infty} \left[(2 + 5n)^2 p^{2n} + (3 + 5n)^2 p^{2n} + 2(5n)^2 p^{2n-1} \right] \quad (13)$$

is generated. Completion of the summation then gives

$$k \equiv \frac{\langle l^2 \rangle}{a^2} = \frac{1}{2}(1 - p) \left[\frac{13}{(1 - p^2)} + \frac{50p^2}{(1 - p^2)^2} + \frac{50p(1 + p^2)}{(1 - p)(1 - p^2)^2} \right] \quad (14)$$

To calculate $\langle s^2 \rangle$ from $\langle l^2 \rangle$, we require the number n of effective bonds. We use the Kuhn approximation [6] $Na = n\langle l^2 \rangle^{1/2}$ to obtain n in terms of N, whence with eq. (10) we find

$$\frac{\langle s^2 \rangle}{Na^2} = \frac{\langle l^2 \rangle}{Na^2} \left(\frac{1 + \cos \alpha}{1 - \cos \alpha} \right) = \frac{k^{1/2}}{2} \quad (15)$$

Equations (11) and (15) afford estimates of $\langle s^2 \rangle / Na^2$ that agree within 2% over the range $0.1 \leq p \leq 1$. We note too that eqs. (11) and (15) give $\langle s^2 \rangle / Na^2$ equal to 1.300 and 1.275, respectively, for the all-meta chain, in excellent agreement with the exact result, eq. (9). Figure 2 is a plot of $\langle s^2 \rangle / Na^2$ as a function of the fraction of meta links that serves for either relation. Comparison with the experimental radii of gyration for fractions of polymer 10-71 indicate that 80% ($\pm 5\%$) of the placements are para. At this level $\langle l^2 \rangle^{1/2}$ is $16.8a$, or 75 Å. It is, of course, this large effective bond length that causes $\langle s^2 \rangle$ to be so large for these relatively low-molecular-weight polymers.

Since these results are unlike model compound data [3] in showing a preference for para catenation, one might question whether $\langle s^2 \rangle / Na^2$ should in fact be given by a relation derived for indefinitely long chains. Some indication of the soundness of this approximation is given by the complete expression [14] for the radius of gyration of a chain with fixed angles $\pi - \alpha$ and links of fixed length l

$$\langle s^2 \rangle = \frac{nl^2}{6} \left(\frac{1 + q}{1 - q} \right) S(n,q) \quad (16)$$

with $q = \cos \alpha$ and

$$S(n,q) = \frac{n + 2}{n + 1} \left\{ 1 - \frac{6q}{(n + 1)(1 - q^2)} \left[1 - \frac{2q}{(n + 1)(1 - q)} \right. \right.$$
$$\left. \left. + \frac{2q^2(1 - q^n)}{n(n + 1)(1 - q)^2} \right] \right\} \quad (17)$$

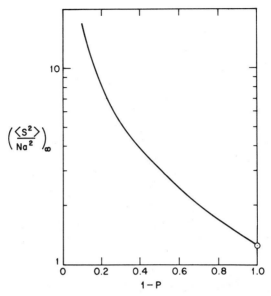

FIG. 2. Characteristic chain dimension for phenyl-substituted polyphenylene calculated from eq. (15) versus the fraction of meta catenation.

It is readily shown [15] that $\langle s^2 \rangle/n$ is within 20% of its limiting value $l^2 (1 + q)/(1 - q)$ if $(1 - q)n/q > 10$. For all the polymers discussed here, this parameter, estimated with $n = N/k^{1/2}$, meets the requirement. Hence the experimental result that $\langle s^2 \rangle/Na^2$ is constant is consistent with eq. (17).

Data on fractions of polymers 9-68 and 10-73 have not been included in the analysis above. Values of $\langle s^2 \rangle_{LS}/N_z a^2$, and even $\langle s^2 \rangle_{LS}/N_w a^2$, for these materials are in the range 1.2–4, much smaller than the value 8.5 found for the fractions of 10-71. Together with the chromatographic elution data, we take this to indicate the presence of metastable aggregates of more compact conformation than the unassociated chains, perhaps resembling randomly branched polymer, and not as evidence of greater meta catenation in these polymers.

The model we have used to rationalize the radius-of-gyration data also serves to explain why the polyphenylene exhibits no depolarization of Rayleigh scattering, even though the optical anisotropy δ_0 of the long effective chain links is probably large. Benoit [13] has calculated the anisotropy δ of a chain with free rotation about fixed valence-bond angles:

$$\delta^2 = \frac{\delta_0^2}{n} \left[\frac{1 + b}{1 - b} - \frac{2b(1 - b^n)}{n(1 - b)^2} \right] \tag{18}$$

where $b = (3q^2 - 1)/2$. It turns out that for practical purposes δ^2/δ_0^2 can be represented very well as a single-valued function of $n(1 - q)/q$. For $n(1 - q)/q > 0$, δ^2 is smaller than δ_0^2 by at least a factor of 10; hence, an inconsequential experimental magnitude of δ^2 is reasonable.

Second Virial Coefficient

Values of the second virial coefficient defined by eq. (1) as half the initial slope of $Kc/R(0,c)$ versus c are given in Table I for fractions of polymer 10-71. The magnitudes are of the order expected for ordinary polymers in good solvents. It will also be noted that A_2^{LS} decreases as molecular weight increases. Qualitatively, this is the behavior expected for solutions in a thermodynamically "good" solvent; but the rate of decrease is much greater than is found with typical flexible-chain species: a dependence roughly on $M^{-1.3}$ here as compared with perhaps $M^{-0.3}$ in extreme cases with vinyl polymers [16]. However, since the polyphenylene chains are short in terms of numbers of effective links, this difference need not, of itself, be cause for surprise. For the same reason, theoretical deductions [17a,18] of the asymptotic dependence of A_2 on molecular weight would hardly seem pertinent.

The other information from light scattering listed in Table I can be combined with the second virial coefficient to obtain the quantity

$$\Psi = A_2^{LS} M_w^{1/2} (\langle s^2 \rangle_{LS} / M_w)^{-3/2} \tag{19}$$

This falls in the range $0.05 \times 10^{24} - 0.08 \times 10^{24}$ for the polyphenylene fractions, in marked contrast to the much larger limiting experimental value $\Psi_\infty = 4.0 \times 10^{24}$ attained for monodisperse, linear, flexible-chain polymers in good solvents [17b,19]. To develop a tentative interpretation for this small Ψ, we recall that interaction between a pair of chain segments is characterized by an excluded volume integral [17c] that has a larger positive value, the "better" the solvent for a given polymer. The net repulsion between segments in a good solvent is manifested by an expansion $\alpha^2 \equiv \langle s^2 \rangle_0$ and by a positive second virial coefficient. The well-known perturbation treatment [17a, 17c] of both effects gives α^2 and A_2 as series expansions, for homogeneous polymer,

$$\alpha^2 = 1 + (134/105)z + O(z^2) \tag{20}$$

and

$$A_2 = 4\pi^{3/2} N_A B [1 - 2.865z + O(z^2)] \tag{21}$$

in powers of the variable

$$z = B(\langle s^2 \rangle_0 / M)^{-3/2} M^{1/2} \tag{22}$$

In these equations N_A denotes Avogadro's number and B is proportional to the excluded volume integral. Eliminating z between the two series, Zimm, Stockmayer, and Fixman [20] obtained A_2 as a function of $(\alpha^2 - 1)$, i.e., in linear approximation

$$A_2 M^{1/2} = 4(105/134)\pi^{3/2} N_A [\langle s^2 \rangle_0 / M]^{3/2} (\alpha^2 - 1) \tag{23}$$

Although it is formally limited to $\alpha^2 \approx 1$, eq. (23) correlates typical experimental data surprisingly well over a wide range of α and A_2. It can be recast in the form

$$\alpha^2 \Psi = 1.05 \times 10^{25} (\alpha^2 - 1) \tag{24}$$

If we accept this relation, ignoring polydispersity as a first approximation, we must conclude that α for the 10-71 fractions is never greater than 1.01. Then the excluded volume parameter z according to eq. (20) is always less than 0.01, and B is simply $A_2/4N_A\pi^{3/2}$. This means that there is virtually no intramolecular excluded volume effect and $\langle s^2 \rangle$ is indistinguishable from the unperturbed dimension, despite a large B, which is expressed in the intermolecular interactions measured by A_2. In other words, the experimental data on the polyphenylenes are explicable on the basis of an intramolecular excluded volume effect that is small because of a very large unperturbed chain dimension, not because of a small excluded-volume integral. This rationale has also been proposed to account for the similar behavior of cellulose acetate and other cellulose esters in solution [21].

Intrinsic Viscosity

The molecular-weight dependence of the intrinsic viscosity of polyphenylene fractions is shown in Figure 3 (some of the data are included in Table I). It is evident that $[\eta]$ is proportional to M_w determined by light scattering for fractions of polymer 10-71, but the other samples deviate markedly from this correlation. If the chain conformation in these polymers is essentially unperturbed, the proportionality between $[\eta]$ and M_w implies that the hydrodynamic behavior approximates the "free draining" limit. For this case, the bead-and-spring chain model gives [6b, 17d]

$$[\eta] = \pi \langle s^2 \rangle d_s / 200 m_s; \qquad d_s = \zeta_0 / 3\pi\eta_0 \qquad (25)$$

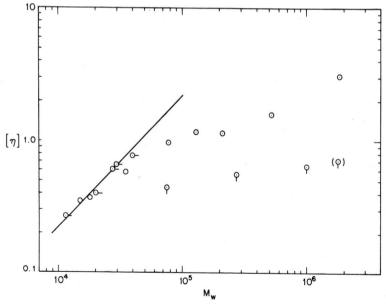

FIG. 3. Intrinsic viscosity versus M_w for polyphenylene fractions. Data points are designated as in Figure 1.

where ζ_0 is the friction factor for an effective hydronamic unit of mass m_s and Stokes diameter d_s, and η_0 is the solvent viscosity. In this model the frictional forces arise at beads of diameter d_s separated by a mean-square distance $b^2 = 6\langle s^2 \rangle / n'$, where n' is the number of links in an equivalent freely-jointed chain with $\langle s^2 \rangle$ matching the value for the real chain. For comparison with experiment we use a form of eq. (25) corrected [19], approximately, for polydispersity:

$$\frac{[\eta]}{M_w} = \frac{\pi d_s}{200 m_s} \left(\frac{\langle s^2 \rangle_{LS}}{M_z} \right) \tag{26}$$

With the Kuhn approximation, $n'b = N_z a$, and the experimental $\langle s^2 \rangle_{LS}$, we obtain $b = 224$ Å, or an averaged link length containing 51 backbone phenylene groups. The experimental parameters $[\eta]/M_w = 2.17 \times 10^{-5}$ and $\langle s^2 \rangle_{LS}/M_z = 9.84 \times 10^{-17}$ then give $d_s = 20$ Å.

The artificiality of this model is revealed in the value of d_s, which cannot be associated with a molecular dimension. Similar difficulties have been encountered in studies of cellulose esters [21] and heterocyclic flexible-chain polymers [22]. Consequently, indiscriminate use of eq. (25) to estimate chain dimensions from viscometric data cannot be advised. We observe, however, that if we arbitrarily take m_s as the mass of the 75-Å effective link for the free rotation model and use the experimental $[\eta]$ and $\langle s^2 \rangle_{LS}$ in eq. (26), we obtain $d_s = 6.8$ Å, a number in plausible correspondence with a chain diameter.

The effects of association, already noted, are evident in the viscosity data on fractions from polymers 9-68 and 8-73. Intrinsic viscosities are much smaller than would be expected by comparison with the "well-behaved" 10-71 series. Moreover, the Mark-Houwink-Sakurada exponent $d \ln [\eta] / d \ln M$ is less than 0.5 for both polymers. These effects are consistent with the presence of aggregates with conformations resembling randomly branched chains [21].

Since exclusion chromatography is carried out at much lower polymer concentration than light scattering and viscometry and any reversible aggregation tends to disappear with decreasing concentration, it is of some interest to obtain a molecular weight M_e from the elution volumes of the 9-68 and 8-73 fractions, assuming that the universal calibration shown in Figure 1 together with the proportionality between $[\eta]$ and M_w shown in Figure 3 holds for these polymers in terms of the molecular weights of species existing in the column. Then, comparing the light scattering M_w with M_e, we could obtain some indication of the minimum degree of association under the conditions of the light-scattering experiments. In this way it is deduced that the aggregates apparently comprise two to five chains, at least, for fractions of polymer 9-68 and from 3 to 36 chains for fractions of 8-73.

Pursuing this line of thought to even more speculative lengths, we can use the M_e calculated as just described and suppose that the aggregates exhibit the viscometric behavior of free-draining branched coils; i.e., the intrinsic viscosity is given by eq. (25) in terms of $\langle s^2 \rangle$ of the aggregate. In this case the degree of association ν can be obtained from

$$[\eta] = 2.17 \times 10^{-5} g \nu M_e \tag{27}$$

where g, the ratio of radii of gyration of branched and linear species of the same mass, is a function of branching topology and the frequency of branch nodes. Using values of g for random tetrafunctional branching given by Zimm and Stockmayer [23] with out experimental viscosities, we find the apparent ν to be in the range 0.5–1 for the fractions of polymers 9-68 and 8-73. Since concentrations are higher in viscometry than in light scattering, it is surprising that this result is contrary to the other, more direct, evidences of association. However, it bears emphasis that the argument by which it was reached is tenuous. It ignores possible heterogeneity of aggregates and would be vitiated if their conformation were far from the random coil or if free-draining behavior did not apply, owing to their enhanced segment density.

Fluorescence

The absorption spectrum of a THF solution of the polyphenylene (0.04 g/liter, 10.05-mm path length) shows broad absorption maxima at 305, 285, and 276 nm, and absorption increasing below 230 nm. Irradiation of a solution (6×10^{-4} g/liter) in THF at 305 nm produces intense fluorescence with a maximum at 365 nm. Excitation at 285 or 276 nm produces only weak fluorescence, but this also peaks at 365 nm. The fluorescence is independent of molecular weight.

The fluorescence emission anisotropy r defined by eq. (1) is a measure of the depolarization of the fluorescence. The experiment has the advantage of simplicity, but it is seriously limited with regard to quantitative interpretation for lack of a satisfactory theory relating r to macromolecular properties. For rigid particles with rotatory diffusion constant D_r, the fluorescence anisotropy is given by [11]

$$r = r_0/(1 + 6D_r\tau) = r_0/(1 + 3\tau/\rho) \tag{28}$$

where τ is the lifetime of the excited state and $\rho = \frac{1}{2}D_r$ is the rotatory relaxation time. The intrinsic emission anisotropy r_0, the limit of r as $D_R\tau$ vanishes, is determined by the angle β between absorption and emission vectors:

$$r_0 = (3 \cos^2 \beta - 1)/5 \tag{29}$$

and thus lies between -0.2 and 0.4. Using the relation

$$D_r = RT/4M\eta_0[\eta] \tag{30}$$

for a polymer chain rotating as a unit, we can put eq. (28) in the form

$$\frac{1}{r} = \frac{1}{r_0} + \frac{3}{200r_0} \left(\frac{\tau RT}{M\eta_0[\eta]} \right) \tag{31}$$

No adequate theory exists to relate r to the properties of chain molecules with a multiplicity of internal normal modes, but it has been suggested [24] that eq. (28) be modified by putting it in a form that typically appears in the analysis of systems characterized by a distribution of relaxation times:

$$r = r_0 \sum \frac{A_i}{1 + 3\tau/\rho_i} \tag{32}$$

The A_i are a normalized set ($\Sigma A_i = 1$) of weighting factors for the modes with relaxation times ρ_i. If every term $3\tau/\rho_i$ and the sum $3\tau\Sigma_i A_i\rho_i$ are small compared with unity, eq. (32) can be simplified [24]:

$$r \approx r_0 \left[2 + 3\tau \sum_i A_i/\rho_i \right]^{-1} \tag{33}$$

Further if the ρ_i are expressed in terms of dimensionless quantities λ_i

$$\rho_i = 100 M \eta_0[\eta]/RT\lambda_i \tag{34}$$

we obtain from eq. (33)

$$\frac{1}{r} = \frac{1}{r_0} + \frac{3\Sigma A_i\lambda_i}{100r_0} \left(\frac{\tau RT}{M\eta_0[\eta]} \right) \tag{35}$$

where $\Sigma A_i\lambda_i$ is just a number that correlates r with the experimental variables. Equation (35) is of the same form as eq. (31) for a rigid body. Thus, a plot of r^{-1} versus $\tau RT/M\eta_0[\eta]$ should extrapolate linearly to an intercept r_0^{-1}, and one might hope to learn something about internal molecular dynamics from slope, i.e., from comparison of the apparent value of $\Sigma A_i \Lambda_i$ with values given by possible theoretical models and with the rigid-body relation in eq. (31).

To carry out such an analysis, we need to know the lifetime τ. For the polyphenylene this was determined to be 4.92 nsec by an independent measurement of fluorescence decay. The emission intensity curve $I(t)$ after excitation by a pulse was fitted by an exponential decay function

$$I(t) = I_0 e^{-t/\tau} \tag{36}$$

in time t. Instrumentation and numerical deconvolution techniques used in this method are described elsewhere [25].

A plot of r^{-1} is shown in Figure 4(a) for fractions of polymer 10-71 in dimethyl phthalate over the temperature range 25–150°. The straight line drawn through the data gives $r_0 = 0.314$ and a slope of 0.653, which corresponds to $\Sigma A_i\lambda_i = 6.85$, according to eq. (35). The slope is much greater than the value 0.047 predicted by eq. (31) with $r_0 = 0.314$. Of course, the failure of the rigid-body model is hardly surprising in view of the flexibility of the polyphenylene chain. Nonetheless, the reasonable correlation of $r^{-1}\tau RT/M_w\eta_0[\eta]$ does suggest that long-range motions of the chain are important, as compared, for example, with side-chain relaxations.

The magnitudes of the experimental fluorescence anisotropy parameters, however, raise further questions about the analysis. The value of r_0 is unexpectedly small. If, as we suppose, the absorption and emission vectors both lie along the axis of the rodlike chain segment, β in eq. (9) is zero, and r_0 should be 0.4. We also note that if $\Sigma A_i = 6.85$, the condition that $3\tau\Sigma A_i/\rho_i \ll 1$ fails when $\tau RT/M\eta_0[\eta]$ is greater than about unity. Consequently, most of the experimental data in Figure 3(a) fall outside the range of the independent variable where the linear relation in eq. (35) can be expected to hold, and the extrapolation to $1/r_0$ may be much more hazardous than appears at first sight.

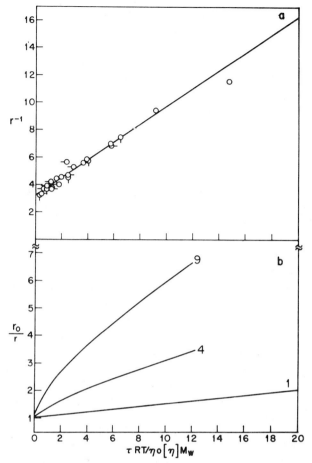

FIG. 4. (a) Reciprocal emission anisotropy r^{-1} for fractions of sample 10-71.
(b) r_0/r calculated with eq. (32) as described in the text for $N = 1, 4, 9$.

To assess this extrapolation hazard, it is instructive to consider an illustrative example of the effect of a distribution of relaxation times. For this purpose we suppose there are N equally weighted modes ($A_i = 1/N$) with $\lambda_i = (\pi^2/6)i^2$, ($i = 1,3, \ldots, N$), corresponding to the N longest relaxation times for the Rouse model [17d]. For this case. eq. (35) becomes

$$\frac{1}{r} = \frac{1}{r_0} + \frac{\pi^2(N + 1)(2N + 1)}{1200 r_0}\left(\frac{\tau RT}{M \eta_0[\eta]}\right) \qquad (37)$$

which may be compared with r^{-1} calculated directly from eq. (32) with eq. (34). Curves from the latter calculation are shown in Figure 4(b) for N equal to 2, 4, and 9. It is evident that when N is 7 or 9, the limiting linear relation in eq. (37) agrees with the curves only when $\tau RT/M\eta[\eta]$ is too small to be experimentally accessible. However, the curvature of the plots is so gentle that experimental points might easily be taken to lie on a straight line. Then, a plausible empirical fitting with the form of eq. (35) would give values of $\Sigma A_i \lambda_i / r_0$ and of r_0 that are too small. As an exercise in curve fitting, the experimental data in Figure 7(a) can be accommodated to eq. (33) with the choice of the λ_i and A_i specified above to give $\rho_0 = 0.4$ with $N \approx 5$. The arbitrary character of the calculation makes it impossible to take this result seriously except to indicate the hazard in using eq. (35) to analyze data on flexible chains.

CONCLUSIONS

Our solution measurements are consistent with polyphenylene chains characterized by rodlike segments of average length 70–75 Å joined by 120° valence angles. The large segment length causes the reduced radius of gyration $\langle s^2 \rangle_0 / M$ to be very large. As a consequence, intramolecular volume exclusion effects on chain conformation are negligible, even though interactions evidenced by the positive second virial coefficient do not vanish. Analogous behavior together with a large $\langle s \rangle_0 / M$ is well known for cellulose derivatives.

The large chain dimension of the polyphenylene is also responsible for free-draining hydrodynamic behavior; i.e., the Mark-Houwink-Sakurada exponent is unity, despite the lack of appreciable intramolecular volume exclusion. Again such behavior is characteristic of cellulose derivatives.

Finally, although fluorescence anisotropy data do suggest the importance of dynamic modes involving concerted motions of extensive parts of the polymer chain, the theoretical development is too crude to permit quantitative interpretation.

On the whole, we must regard our conclusions tentative rather than definitive for at least three reasons. (1) The data are limited to a small number of available samples. (2) The statistical theories interrelating conformational, hydrodynamic, and thermodynamic behavior are all for chains comprising large numbers of statistical segments, but the polyphenylene chains are quite short in this sense. (3) Since the problem of intermolecular association frustrated interpretation of data on some polymer preparations, the assumption that aggregates are completely absent in the "well-behaved" samples is obviously problematical.

This study was supported by Grants GP-28538X and DMR74-14953 of the National Science Foundation. JLW acknowledges the support of the Armstrong Cork Company, which made possible his appointment as a Visiting Fellow at Carnegie-Mellon University. Dr. L. K. Patterson obtained the data on fluorescence decay. Assistance on certain phases of the study was rendered by R. W. Nelb and Dr. G. K. Noren, of the University of Iowa, and by S. M. Liwak.

REFERENCES

[1] G. K. Noren and J. K. Stille, *J. Polym. Sci. Part D, 5,* 385 (1971).

[2] *J. Polym. Sci. Part B, 7,* 525 (1969); J. K. Stille and G. K. Noren, *Macromolecules, 5,* 49 (1972).

[3] H. Mukamal, F. W. Harris, and J. K. Stille, *J. Polym. Sci. Part A-1, 5,* 2721 (1967).

[4] Z. Grubisic, P. Rempp, and H.Benoit, *J. Polym. Sci. Part B, 5,* 753 (1967).

[5] G. C. Berry, *J. Polym. Sci. Part A-2, 9,* 687 (1971).

[6] P. J. Flory, "Principles of Polymer Chemistry," Cornell University Press, Ithaca, 1953, (a) Chap. 7; (b) Chap. 14.

[7] G. C. Berry, *J. Chem. Phys., 44,* 4550 (1966).

[8] B. H. Zimm, *J. Chem. Phys., 16,* 1099 (1948).

[9] E. F. Casassa and G. C. Berry in "Polymer Molecular Weights," P. E. Slade, Jr., Ed., Dekker, New York, 1975, Chap. 5.

[10] G. C. Berry, *J. Chem. Phys., 46,* 1338 (1967).

[11] D. J. R. Laurence, in "Physical Methods in Macromolecular Chemistry," B. Carroll, Ed., Dekker, New York, 1969, Chap. 5.

[12] M. V. Volkenstein, "Configurational Statistics of Polymer Chains," Engl. Trans. by S. N. Timasheff and M. J. Timasheff, Wiley-Interscience, New York, 1963, p. 195.

[13] H. Benoit, C. R. *Acad. Sci., 236,* 687 (1953).

[14] H. Benoit and P. M. Doty, *J. Phys. Chem., 57,* 958 (1953); R. A. Sack, *Nature, 171,* 310 (1953).

[15] G. C. Berry, *J. Polym. Sci. Polym. Symp.*

[16] E. F. Casassa, *Polymer, 3,* 621 (1962).

[17] H. Yamakawa, "Modern Theory of Polymer Solutions," Harper and Row, New York, 1971, (a) Chap. 4; (b) Chap. 7; (c) Chap. 3; (d) Chap. 6.

[18] M. Daoud, J. P. Cotton, et al., *Macromolecules, 8,* 804 (1975).

[19] G. C. Berry and E. F. Casassa, *J. Polym. Sci. Part D, 4,* 1 (1970).

[20] B. H. Zimm, W. H. Stockmayer, and M. Fixman, *J. Chem. Phys., 21,* 1716 (1953).

[21] D. W. Tanner and G. C. Berry, *J. Polym. Sci. Polym. Phys. Ed., 12,* 941 (1974).

[22] G. C. Berry, *Discuss. Faraday Soc., 49,* 121 (1970).

[23] B. H. Zimm and W. H. Stockmayer, *J. Chem. Phys., 17,* 1302 (1949).

[24] M. Frey, P. Wahl, and H. Benoit, *J. Chim. Phys., 61,* 1005 (1964).

[25] L. K. Patterson and E. Vieil, *J. Phys. Chem., 77,* 1191 (1973).

PROPERTIES OF AN OPTICALLY ANISOTROPIC HETEROCYCLIC LADDER POLYMER (BBL) IN DILUTE SOLUTION

G. C. BERRY

Department of Chemistry, Carnegie-Mellon University, Pittsburgh, Pennsylvania 15213

SYNOPSIS

Light scattering, intrinsic viscosity, gel-permeation chromatography, and fluorescence depolarization measurements are reported on dilute solutions of a heterocyclic polymer (BBL) prepared by polymerization of 1,4,5,8-naphthalene tetracarboxylic acid with 1,2,4,5-tetraaminobenzene. The light-scattering data on fractions of the polymer giving the anisotropy δ and the mean-square radius of gyration $\langle s^2 \rangle$ are discussed in terms of equations for the scattering from polydisperse anisotropic polymer. It is concluded from the dependence of δ and $\langle s^2 \rangle$ on the molecular weight that the polymer has a nearly rodlike conformation. The intrinsic viscosity data are analyzed in terms of relations for rods and wormlike chains. The fluorescence emission anisotropy is found to be so sensitive to molecular-weight distribution that analysis in terms of molecular parameters is prohibited. Association that can occur under certain circumstances is discussed.

INTRODUCTION

In previous communications we discussed the properties of the heterocyclic polymer poly(6,9-dihydro-6,9-dioxobisbenzimidazo[2,1-*b*:1′,2′-*j*]-benzo-[1mn][3,8] phenanthroline-3,12-diyl), referred to as BBB, prepared by polymerization [1] of 3,3′-diaminobenzidine with 1,4,5,8-naphthalene tetracarboxylic acid [2–5]. Here, we report investigations on dilute solutions of a related polymer poly[(7-oxo-7,10-*H*-benz[de]-imidazo[4′,5′:5,6]benzimidazo[2,1-*a*] isoquinoline-3,4:10,11-tetrayl)-10-carbonyl], referred to as BBL, that is obtained if 1,2,4,5-tetraaminobenzene is used in place of 3,3′-diaminobenzidine in the polymerization [6]. With BBL, an inflexible ladder polymer is obtained if intramolecular cyclization is complete, whereas BBB polymer contains single bonds in the chain backbone that allow rotational isomerization, giving the polymer flexible coil properties [2]. Similar to the situation with BBB, there are few solvents for BBL. A major class of solvents are those which can protonate the polymer [4], such as sulphuric acid, methane sulfonic acid, and benzene sulfonic acid. In this study we investigate the anisotropic light scattering of dilute solutions of BBL in methane sulfonic acid, together with the dependence of the intrinsic viscosity and the fluorescence emission anisotropy on molecular parameters.

Journal of Polymer Science: Polymer Symposium 65, 143–172 (1978)
© 1978 John Wiley & Sons, Inc.   0360-8905/78/0065-0143$01.00

THEORETICAL CONSIDERATIONS

Light Scattering from Anisotropic Polymers

Equations needed to analyze light-scattering data on anisotropic polymers are brought together in this section and extended to include effects of molecular-weight heterogeneity. The light scattering from a polymer chain comprised of anisotropic elements involves the intrinsic anisotropy factor δ_0 defined in terms of the principal polarizabilities α_1, α_2, and α_3 of the scattering element [7, 8]:

$$\delta_0^2 = \frac{(\alpha_1 - \alpha_2)^2 + (\alpha_1 - \alpha_3)^2 + (\alpha_2 - \alpha_3)^2}{2(\alpha_1 + \alpha_2 + \alpha_3)^2} \tag{1}$$

(The reader should be aware that alternative definitions of a quantity designated δ_0 are sometimes used [9–10].) For the case of scattering elements with cylindrical symmetry with $\alpha_2 = \alpha_3 = \beta$ and $\alpha_1 = \alpha$, this reduces to the result

$$\delta_0 = \frac{\alpha - \beta}{\alpha + 2\beta} \tag{2}$$

In this case, δ_0 lies in the range $-1/2 < \delta_0 < 1$. The overall anisotropy δ for a polymer chain can be defined as an average of δ_0 for each chain segment over conformation space.

It is convenient at this point to consider two chain models that have been used in discussions of the light scattering from anisotropic polymers. These are (1) a chain with free rotation about each of the N bonds with length l, anisotropy δ_0, and valence angle $\pi - \alpha$ and (2) a wormlike chain with n equivalent segments of length b, anisotropy δ_0, and a persistence length $\rho = b/2\lambda$, where λ is a parameter between zero and unity. The wormlike chain includes the rod and random-coil models in the limits as λ goes to zero and unity, respectively. The overall chain anisotropy δ has been calculated with both models, giving the results [11, 12]

$$\delta^2 = \frac{\delta_0^2}{N} \left\{ \frac{1 + b}{1 - b} - \frac{2b}{N} \frac{1 - b^N}{(1 - b)^2} \right\} \tag{3}$$

where

$$b = \frac{3a^2 - 1}{2}$$

$$a = \cos \alpha$$

for the chain with free rotation and

$$\delta^2 = \delta_0^2 \left(\frac{2}{3Z} \right) \{1 - (3Z)^{-1}[1 - \exp(-3Z)]\} \tag{4}$$

where $Z = 2\lambda n = L/\rho$ for the wormlike chain, with $L = nb$ the contour length of the chain. The ratio $(\delta/\delta_0)^2$ given by eq. (3) is shown as a function of $N(1 - a)/a$ in Figure 1 for several values of α. The parallel behavior given by

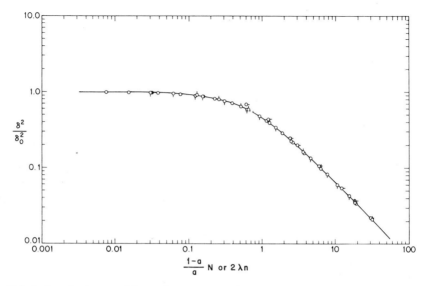

FIG. 1. Overall anisotropy δ^2 for a polymer chain with free rotation about bonds with valence angles $(\pi - \alpha)$ with $a = \cos \alpha$ and N the number of bonds. The points are calculated with $\alpha = 5°$, O; $10°$, O; $20°$, ♀; and $40°$, ○. The solid curve gives δ^2/δ_0 versus $2\lambda n$ according to eq. (4).

eq. (4) is also shown in Figure 5. Equation 4 provides a very good representation of eq. (3) if the Z is put equal to $N(1 - a)/a$ in the former. With either representation, δ^2 is proportional to M^{-1} in the limit of large molecular weight M.

The mean-square radius of gyration $\langle s^2 \rangle$ has also been calculated with both models, with the result [13, 14a]

$$\langle s^2 \rangle = \frac{Nl^2}{6} \frac{1 + a}{1 - a} S(N,a) \tag{5a}$$

$$S(N,a) = \left(\frac{N + 2}{N + 1}\right) \left\{ 1 - \frac{6a}{(N + 2)(1 - a^2)} \right.$$
$$\times \left[1 - \frac{2a}{(N + 1)(1 - a)} + \frac{2a^2(1 - a^N)}{N(N + 1)(1 - a)^2} \right] \right\} \tag{5b}$$

and

$$\langle s^2 \rangle = \frac{L\rho}{3} S(Z) \tag{6a}$$

$$S(Z) = 1 - 3Z^{-1} + 6Z^{-2} - 6Z^{-3} [1 - \exp(-Z)] \tag{6b}$$

Equations (5) and (6) give limiting values $(\langle s^2 \rangle / N)_\infty$ of $\langle s^2 \rangle / N$ at large M of $(l^2/6)(1 + a)/(1 - a)$ and $L\rho/3N$, respectively. The function $S(N,a)$ and $S(Z)$, shown in Figure 2, are in close agreement with $Z = N(1 - a)/a$, except for very small N.

The special cases of random coil and rodlike chain models are both encompassed as limiting cases in eqs. (5) and (6). The former, with $\langle s^2 \rangle = nb^2/6$ and $\delta^2 = \delta_0^2/n$, is recovered in the limit as L/ρ goes to infinity, whereas the latter, with $\langle s^2 \rangle = L^2/12$ and $\delta = \delta_0$, is recovered in the limit as L/ρ goes to zero.

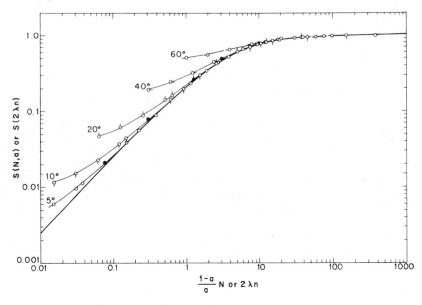

FIG. 2. The function $S(N,a)$ for a chain with free rotation about bonds with valence angles $(\pi - \alpha)$ with $a = \cos \alpha$ for the indicated α. The heavy curve is $S(2n\lambda)$ versus $2\lambda N$ according to eq. (6).

The light scattering from a polymer with anisotropic elements is given in terms of the Rayleigh ratios R_{Vv} and R_{Hv} for the vertical and horizontal components of light scattered in the transverse plane with vertically polarized incident radiation, respectively, and the similar quantities R_{Hh} and R_{Vh} for horizontally polarized incident light. For molecules of interest here, $R_{Vh} = R_{Hv}$. Relations in the literature for rods, random coils, coils with free rotation and fixed valence angles, and wormlike chains can be expressed compactly in terms of the overall anisotropy δ and the parameter

$$u = \langle s^2 \rangle h^2$$

where $h = (4\pi n/\lambda_0) \sin(\theta/2)$. Thus, to order u,

$$\left[\frac{R_{Vv}(u,\delta)}{KMc} \right]^0 = \left(1 + \frac{4}{5}\delta^2 \right) - \frac{1}{3}\left[1 - \frac{4}{5}f_1\delta + \frac{4}{7}(f_2\delta)^2 \right] u + 0(u^2) \quad (7)$$

$$\left[\frac{R_{Hv}(u,\delta)}{KMc} \right]^0 = \frac{3}{5}\delta^2 - \frac{9}{35}(f_3\delta)^2 u + 0(u^2) \quad (8)$$

$$\left[\frac{R_{Hh}(u,\delta)}{KMc} \right]^0 = \frac{3}{5}\delta^2 + \cos^2\theta\left[1 + \frac{1}{5}\delta^2 \right]$$
$$- \left\{ \left[\frac{2}{5}f_1\delta + \frac{13}{35}(f_4\delta)^2 \right] + \frac{\cos^2\theta}{3}\left[1 + \frac{2}{5}f_1\delta + \frac{1}{7}(f_2\delta)^2 \right] \right\} u + 0(u^2) \quad (9)$$

where the functions f_1, f_2, f_3, and f_4, all lying between unity and zero, depend on the chain model. The superscript 0 denotes the scattering at infinite dilution, c the concentration, and K an optical constant.

All the f_i are unity for a rodlike chain [7] and zero for a random coil [8]. They all depend on a and N for the chain with free rotation [15], or on L/ρ for wormlike chains [12]. Relations for R_{V_v} to order u given by Benoit [15] and Nagai [12] for the freely rotating chain and the wormlike chain, respectively, can be used to calculate f_1 for these models. The results, given in Figure 3, show that the two models give equivalent results using the correspondence $Z = N$ $(1 - a)/a$ noted above. Nagai's expressions for R_{Vh} and R_{Hh} can be used to calculate the functions f_2, f_3, and f_4 for the wormlike chain, with the results shown in Figure 3—we are unaware of comparable calculations for the freely rotating chain model.

The interesting feature of the data in Figure 3 to be noted is that all the f_i decrease fairly slowly with increasing L/ρ. In particular, coefficients such as the factor $T_1 = [1 - (4/s)f_1\delta + (4/7)(f_2\delta)^2]$ in eq. (7) can usually be approximated satisfactorily by their values for rodlike molecules, i.e., with $f_i = 1$. For example, for L/ρ such that f_1 and f_2 differ much from unity, the value of δ is so small that T_1 is nearly unity. With T_1, the maximum error in T_1 so committed is about 12% for the case $\delta_0 = 1$ and $L/\rho = 0.5$ (or $\delta^2 = 0.1$). Consequently, in the application of eqs. (7)–(9) for analysis of anisotropic light-scattering data, one can usually put all the f_i equal to unity with small to negligible error, even for a flexible coil polymer.

Evaluation of $\langle s^2 \rangle$, δ, and M from experimental data is not a trivial matter, even if u is small enough to permit application of eqs. (7)–(9). In principle, of course, δ can be determined if both R_{Hv} and R_{Vv} can be extrapolated to zero scattering angle with sufficient accuracy—see eq. (35) below. This is often difficult, owing to anomalous scattering at small angles from extraneous matter or, for example, small amounts of associated polymeric species. Consequently, it is useful to investigate methods to extract δ from data at, say, 90° scattering angle. Equations (7)–(9) may not be generally useful for this purpose, since they are limited to small u. We use instead the relations for R_{Vv}, etc., given by Horn [7] for anisotropic rodlike chains to aid in this analysis. This approximation is

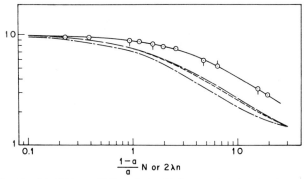

FIG. 3. The function $f_1(a,N)$ in eq. (7) for a chain with free rotation about bonds with valence angle $\pi - \alpha$ with $a = \cos \alpha$ is shown as the circles. The symbols are defined in the caption to Figure 2. The functions $f_1(\lambda n)$, —; $f_2(\lambda n)$, - - -; $f_3(\lambda n)$, - -, in eqs. (7)–(9) for wormlike chain are also shown.

introduced on the premise that unless the chain studied has a nearly rodlike conformation anyway, δ is so small that the particular method used in its determination leads to an estimate adequate for the evaluation of $\langle s^2 \rangle$ and M. Of course, if the conformation is not rodlike, the estimate for δ itself could be in error. After formulating Horn's results in a compact form useful for our further purpose, we extend them to include the effects of molecular-weight heterogeneity and discuss procedures for determining δ on the basis of the equations so obtained.

Horn's results are conveniently given in terms of four functions $q_1(u)$, $q_2(u)$, $q_3(u)$, and $q_4(u)$ that go to unity in the limit as u goes to zero and three functions $m_1(u)$, $m_2(u)$, and $m_3(u)$ that go to zero in the same limit. These, in turn are linear combinations of the three functions

$$p_1(x) = \frac{2[1 - x\,\mathrm{Si}(x) - \cos x]}{x^2} \tag{10}$$

$$p_2(x) = \frac{2[x - \sin x]}{x^3} \tag{11}$$

$$p_3(x) = \frac{12[x^3 + 3x \cos x - \sin x]}{x^5} \tag{12}$$

where $x^2 = 12u$ and $\mathrm{Si}(\ )$ is the sine integral. Then

$$q_1 = p_1 = 1 - \frac{1}{3}u + \cdots$$

$$q_2 = 3p_2 = 1 - \frac{3}{5}u + \cdots$$

$$q_3 = \frac{15}{8}\left[p_2 + \frac{3}{2}p_3\right] = 1 - \frac{15}{28}u + \cdots$$

$$q_4 = 6q_3 - 5q_2 = 1 - \frac{3}{14}u + \cdots$$

$$m_1 = q_1 - q_2 = \frac{4}{15}u + \cdots$$

$$m_2 = p_1 - p_2 - 5p_3 = \frac{16}{105}u + \cdots$$

$$m_3 = p_1 + \frac{5}{3}p_2 - \frac{35}{3}p_3 = 0(u^2)$$

With these definitions, the scattering functions for anisotropic rods given by Horn can be cast in the form, correcting misprints in reference 7,

$$\left(\frac{R_{Vv}(u,\delta)}{KMc}\right)^0_{\mathrm{ROD}} = q_1(u) + \frac{4}{5}\delta^2 q_4(u)$$

$$+ (\delta - 2\delta^2)m_1(u) + \frac{27}{8}\delta^2 m_2(u) \tag{13}$$

$$\left(\frac{R_{Hv}(u,\delta,L/\lambda)}{KMc}\right)^0_{\text{ROD}} = \delta^2 \left\{\frac{2}{5} q_3(u) + \frac{1}{5} q_4(u) + \frac{9}{8} \frac{u}{k} m_2(u)\right\} \quad (14)$$

$$\left(\frac{R_{Hh}(u,\delta,L/\lambda)}{KMc}\right)^0_{\text{ROD}} = \delta^2 \left\{\frac{3}{5} q_3(u) + \frac{63}{64} m_2(u) + \frac{27}{64} m_3(u) - \frac{9}{8} \frac{u}{k} m_2(u)\right\}$$

$$- \frac{6\delta + 3\delta^2}{4} m_1(u) \left(1 - \frac{2u}{k}\right)$$

$$+ \cos^2\theta \left\{q_1(u) + \frac{1}{5}\delta^2 q_4(u) + \frac{27}{32}\delta^2 m_2(u) - \frac{\delta + \delta^2}{2} m_1(u)\right\} \quad (15)$$

and $R_{Vh}(u,\delta,L/\lambda) = R_{Hv}(u,\delta,L/\lambda)$. Here $k = 16\pi^2(\langle s^2\rangle/\lambda^2) = (4\lambda^2/3)$ $(L/\lambda)^2$, where L is the rod length and $\lambda = \lambda_0/n$. It should be noted that R_{Hv} and R_{Hh} depend explicitly on L/λ, whereas R_{Vv} does not. Also, it may be noted that R_{Vv} reduces to the function p_1 if the rod is optically isotropic and that eqs. (13)–(15) reduce to eqs. (7)–(9) to order u.

Scattering Equations for Polydisperse Anisotropic Rods

Since only linear combinations of p_1, p_2, and p_3 appear in eqs. (13)–(15), the effects of molecular-weight heterogeneity on the scattering can be determined by calculating the appropriate "light-scattering" average of the p functions:

$$p_i(x)_{\text{LS}} = M_w^{-1} \int_0^\infty M p_i(x) f(M) \, dM \quad (16)$$

For this purpose, we adopt the familiar Zimm-Schulz distribution function for the weight fraction $f(M) \, dM$ of chains with molecular weight M to $M + dM$ [16]:

$$f(M) \, dM = \frac{(h+1)^{h+1}}{\Gamma(h+1)} \left(\frac{M}{M_w}\right)^h \exp\left(-(h+1)\frac{M}{M_w}\right) \frac{dM}{M_w} \quad (17)$$

with

$$h^{-1} = \frac{M_w}{M_n} - 1 \quad (18)$$

Since, for a rod, $\langle s^2\rangle \, \alpha M^2$, we have $x\alpha M$, and eqs. (16) and (17) can be rewritten in the form

$$p_i(x)_{\text{LS}} = \frac{(h+1)^{h+1}}{\Gamma(h+1)} \int_0^\infty p_i(x) \left(\frac{x}{x_w}\right)^{h+1}$$

$$\times \exp\left(-(h+1)\frac{x}{x_w}\right) d\left(\frac{x}{x_w}\right) \quad (19)$$

where x_w is the weight-average value of x.

The average $p_1(x)_{\text{LS}}$ given by eqs. (10) and (16) has been discussed by Goldstein [17]. For $h > 0$ integration gives

$$p_1(x)_{\text{LS}} = \frac{2}{(h+1)\theta} \left\{\arctan\theta + \sum_{j=1}^{h-1} \left(\frac{1}{h-j} - \frac{1}{h}\right)(1+\theta^2)_x^{(j-h)/2}\right.$$

$$\times \sin \left[(h - j) \arctan \theta\right] \bigg\} \quad (20)$$

with

$$\theta = \frac{x_w}{h + 1}$$

The relation for $p_2(x)_{LS}$ can also be integrated for $h > 0$, with the result

$$p_2(x)_{LS} = \frac{2}{h(h + 1)\theta^2} \left\{1 - (1 + \theta^2)^{(1-h)/2} \frac{\sin \left[(h - 1) \arctan \theta\right]}{(h - 1)\theta}\right\} \quad (21a)$$

which reduces to

$$p_2(x)_{LS} = \frac{1}{\theta^2} \left\{1 - \frac{\arctan \theta}{\theta}\right\} \quad (21b)$$

for $h = 1$. The relation for $p_3(x)_{LS}$ can be integrated analytically in closed form for $h \geq 2$ to give the result

$$p_3(x)_{LS} = \frac{4}{3h(h + 1)\theta^2}$$

$$\times \left\{1 + \frac{3}{(h - 1)(h - 2)\theta^2} (1 + \theta^2)^{(2-h)/2} \left[\cos \left[(h - 2) \arctan \theta\right]\right.\right.$$

$$\left.\left. - (1 + \theta^2) \frac{\sin \left[(h - 3) \arctan \theta\right]}{(h - 3)\theta}\right]\right\} \quad h > 2 \quad (22a)$$

$$p_3(x)_{LS} = \frac{2}{9\theta^2} \left\{1 - \frac{3}{\theta^2}\left[1 - \frac{\arctan \theta}{\theta}\right]\right\} \quad h = 2 \quad (22b)$$

The integral for $p_3(x)_{LS}$ is convergent when $0 < h < 2$ but does not appear to reduce to an elementary function.

The functions $p_1(x)_{LS}$, $p_2(x)_{LS}$, and $p_3(x)_{LS}$ calculated with several values of h are shown in Figure 4. In each case the data are plotted against the parameter $S^2 \equiv \left[(h + 3)(h + 2)/(h + 1)^2\right](x_w/2)^2$, which appears in the expansions

$$p_1^{-1}(x)_{LS} = 1 + \frac{1}{9} S^2 + \cdots \quad (23)$$

$$p_2^{-1}(x)_{LS} = 3 \left(1 + \frac{1}{5} S^2 + \cdots\right) \quad (24)$$

$$p_3^{-1}(x)_{LS} = \frac{15}{2} \left(1 + \frac{1}{7} S^2 + \cdots\right) \quad (25)$$

Inspection of Figure 4 shows that molecular-weight polydispersity results in functions for $1/3p_2(x)_{LS}$ and $2/15p_3(x)_{LS}$ that are more nearly linear in S^2 than is the case with the corresponding functions for monodisperse chains. On the other hand, polydispersity causes $p_1^{-1}(x)_{LS}$ to deviate markedly from linearity with S^2, and, in fact, $p_1^{-2}(x)_{LS}$ is more nearly linear in S^2 than is $p_1^{-1}(x)_{LS}$, as

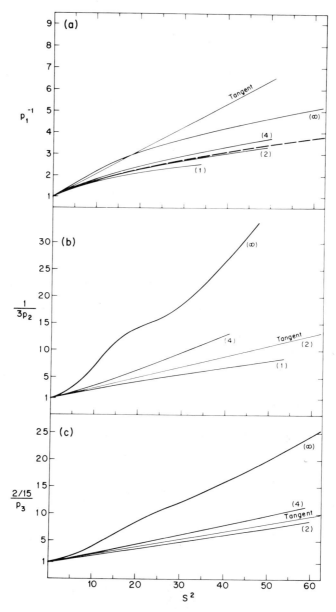

FIG. 4. The primitive scattering functions p_1, p_2, and p_3 as functions of the parameter $S^2 = [(h+3)(h+2)^{-2}](x_w/2)^2$ for the indicated value of the polydisperity factor h. The function $(1 + 2S^2/9)^{1/2}$ is shown in (a) as the dashed line for comparison with p_1^{-1}.

can be seen by comparison of $p_1^{-1}(x)_{LS}$ with the function $(1 + 2/9\,S^2)^{1/2}$ shown in Figure 4.

The scattering functions for polydisperse optically anisotropic rods are given in terms of the average p functions by eqs. (13)–(15), with M_w in place of M, and the definitions for the q and m functions given above.

It is convenient to discuss $R_{V_v}(u,\delta)$ in terms of the reciprocal scattering function $P_{V_v}^{-1}(u,\delta)$ defined by the equation

$$\left[\frac{R_{V_v}(u,\delta)}{KM_wc}\right]^0 = \left(1 + \frac{4}{5}\delta^2\right)P_{V_v}(u,\delta) \qquad (26)$$

The reciprocal scattering function is shown in Figure 5 for h equal to 2, 4 and ∞, the monodisperse case, with δ^2 equal to 0, 0.1, and 1 as a function of the parameter

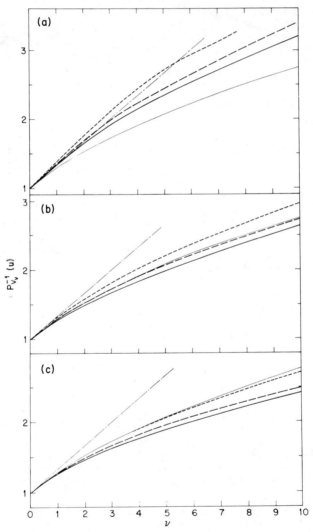

FIG. 5. Reciprocal scattering factor $P_{V_v}^{-1}$ for anisotropic polydisperse rods versus the parameter ν as defined in the text. The curves for (a), $h = \infty$; (b), $h = 4$; and (c), $h = 2$, with $\delta^2 = 0$, - - -; 0.1, - -; and 1.0, —. The light curve is the function $(1 + 2\nu/3)^{1/2}$ for comparison with $P_{V_v}^{-1}$.

$$\nu \equiv \frac{(h + 3)(h + 2)}{(h + 1)^2} \frac{\left(1 - \frac{4}{5}\delta + \frac{20}{35}\delta^2\right)}{\left(1 + \frac{4}{5}\delta^2\right)} u_w$$

which appears in the expansion of $P_{Vc}^{-1}(u)$ for small u_w:

$$P_{Vc}^{-1}(\nu) = 1 + \frac{1}{3}\nu + \cdots \tag{27}$$

It can be seen that polydispersity induces increased downward curvature in $P_{Vc}^{-1}(u)$ and, in fact, would make it impossible to determine $\langle s^2 \rangle_{LS}$ from the initial slope of P_{Vc}^{-1} versus $\sin^2 (\theta/2)$ if experimental data were confined to angles such that $P_{Vc}^{-1} > 1.2$. By contrast, as may be seen in Figure 5, $P_{Vc}^{-1}(\nu)$ is fairly well approximated by the simple relation $(1 + 2\nu/3)^{1/2}$. Thus, if $P_{Vc}^{-1} > 1.2$, it is preferable to plot $[(Kc/R_{Vc}(u,\delta)]^2$ versus $\sin^2 (\theta/2)$ to analyze light-scattering data on polydisperse rods rather than use the more usual plot of $Kc/R_{Vc}(u,\delta)$ versus $\sin^2 (\theta/2)$.

If x_w is very large, the asymptotic form of the reciprocal scattering function is of interest. To order x_x^{-2} the p functions have asymptotic limits given by

$$\lim_{x_w \to \infty} p_1(x)_{LS} = \frac{\pi}{x_w} - \frac{4(h + 1)}{h} \frac{1}{x_w^2} + \cdots \tag{28}$$

$$\lim_{x_w \to \infty} p_2(x)_{LS} = \frac{2(h + 1)}{h} \frac{1}{x_w^2} + \cdots \tag{29}$$

$$\lim_{x_w \to \infty} p_3(x)_{LS} = \frac{4(h + 1)}{3h} \frac{1}{x_w^2} + \cdots \tag{30}$$

These relations may be used to compute the asymptotic form of the q and m functions and hence of the reciprocal scattering curves. For example, $KM_wc/R_{Vc}(u,\delta)$ is given by

$$\lim_{u \to \infty} \left[\frac{KM_wc}{R_{Vc}(u,\delta)}\right]^0 = \frac{2}{\pi^2} \frac{h + 1}{h} \frac{1 + 3\delta - \frac{11}{2}\delta^2}{1 + \delta + \frac{11}{8}\delta^2}$$

$$+ \frac{2}{\pi} \frac{3}{1 + \delta + \frac{11}{8}\delta^2} (u_w)^{1/2} + 0(u_w^{-1}) \tag{31}$$

showing that KM_wc/R_{Vc} tends toward linearity with $u_w^{1/2}$ or, since u_w is large where eq. (31) is applicable, $(K_{Mw}c/R_{Vc})^2$ tends toward proportionality with $(4\pi n/\lambda_0)^2 \sin^2 \theta/2$, with slope $12 \langle s^2 \rangle_{LS}/\pi^2(1 + \delta + 11/8 \delta^2)^2$.

The R_{Uv} scattering given by $R_{Vc} + R_{Hc}$ is more difficult to analyze, since, in general, R_{Uv} is an explicit function of L/λ as well as x_w, or u_w—of course, R_{Uv} can be represented as a function of u_w alone in the limit as L/λ becomes very large. The reciprocal scattering function P_{Uv}^{-1} defined by the relation

$$\left[\frac{R_{Uv}(u,\delta,L/\lambda)}{KM_wc}\right]^0 = \left(1 + \frac{7}{5}\delta^2\right) P_{Uv}\left(u,\delta,\frac{L}{\lambda}\right) \tag{32}$$

can be expanded as an explicit function of

$$\nu' = \frac{(h+3)(h+2)}{(h+1)^2} \frac{1 - \frac{4}{5}\delta + \frac{47}{35}\delta^2}{1 + \frac{7}{5}\delta^2} u_w \tag{33}$$

alone to terms linear ν':

$$P_{Uc}^{-1}\left(\nu',\delta,\frac{L}{\lambda}\right) = 1 + \frac{1}{3}\nu' + \cdots \tag{34}$$

(The first explicit appearance of L/λ appears in the term quadratic in ν'.)

Plots of $P_{Uc}^{-1}(\nu',\delta,L/\lambda)$ are shown in Figure 6 for several values of L/λ, h, and δ. As with $P_{Uc}^{-1}(\nu)$, it is seen that $P_{Uc}(\nu',\delta,L/\lambda)$ displays appreciable downward curvature, making estimation of the initial slope impossible if data are confined to the range $P_{Vc}^{-1} > 1.2$. As can be seen in Figure 6, the function $(1 + 2\nu'/3)^{1/2}$ more nearly fits $P_{Uc}^{-1}(\nu',\delta,L/\lambda)$, although, of course, it lacks the explicit dependence on L/λ that is required for a complete fit to $P_{Uc}^{-1}(\nu',\delta,L/\lambda)$. Nonetheless, plots of $(Kc/R_{Uc}(\theta))^2$ versus $\sin^2 \theta/2$ appear to provide a reasonable basis for analysis of data on $R_{Uc}(\theta)$ if L/λ is large enough and will be used below to analyze data on BBL where appropriate.

Determination of the Chain Anisotropy δ

Analysis of light-scattering data with equations given above requires an estimate for δ. Although this may be readily determined given the ratio R_{Hv}/R_{Vv} extrapolated to zero angle:

$$\left(\frac{R_{Hv}(0,\delta,L/\lambda)}{R_{Vv}(0,\delta)}\right)^0 = \frac{3\delta^2}{5 + 4\delta^2} \tag{35}$$

such an extrapolation is not experimentally easy. Unfortunately, δ is not simply determined from the usual experimental data for either $\rho_U(90°) = R_{Uh}(90°)/R_{Uc}(90°)$ or $\rho_V(90°) = R_{Vh}(90°)/R_{Vc}(90°)$. These may be related to δ by the equations

$$\frac{3\delta^2}{5 + 4\delta^2} = \rho_V(90°)\, Q_V^{-1}[\delta,P_V(90°)] \tag{36}$$

and

$$\frac{6\delta^2}{5 + 7\delta^2} = \rho_U(90°)\, Q_U^{-1}[\delta,P_U^{-1}(90°)] \tag{37}$$

where the functions Q_V and Q_U depend on δ and the reciprocal scattering function at 90° scattering angle. For example, eqs. (7)–(9) may be used to estimate Q_V and Q_U for small u, e.g., for $P(90)$ near unity, with the results

$$Q_V[\delta,P_V^{-1}(90°)] = 1 - b(\delta)[P_V^{-1}(90°) - 1] + \cdots \tag{38a}$$

$$1 + b(\delta) = \frac{6}{7} \frac{1 + \frac{4}{5}\delta^2}{1 - \frac{4}{5}\delta + \frac{4}{7}}\delta^2 \tag{38b}$$

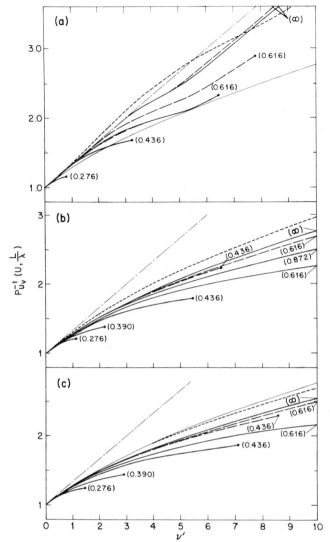

FIG. 6. Reciprocal scattering factor P_{Uc}^{-1} for anisotropic polydisperse rods versus the parameter ν' defined in the text. The curves are for (a) $h = \infty$; (b) $h = 4$; and (c) $h = 2$. In (a) δ^2 is 0, ---; 0.3, – –; and 1.0, —. In (b) and (c), δ^2 is 0, ----; 0.1, – –; and 1.0 —. Values of L/λ are given in parentheses. The light curve in each figure is the function $(1 + 2\nu'/3)^{1/2}$. The initial tangent is shown as the chain line — - —.

and

$$Q_U[\delta, P_U^{-1}(90°)] = 1 - a(\delta)[P_U^{-1}(90°) - 1] + \cdots \qquad (39a)$$

$$1 + a(\delta) = \frac{45}{42} \frac{1 + \dfrac{7}{5}\delta^2}{1 - \dfrac{4}{5}\delta + \dfrac{38}{35}\delta^2} \qquad (39b)$$

The use of eqs. (13)–(15) permits calculation of Q_U and Q_V for larger $P^{-1}(90)$, with the results shown in Figure 7. Although the latter applies strictly only to rodlike chains, we will adopt its use generally on the premise stated above that this approximation should usually give values of δ satisfactory in the estimate of $\langle s^2 \rangle$ and M from light-scattering data.

Given $\rho_U(90°)$ and $P_U(90°)$, or $\rho_V(90°)$ and $P_V(90°)$, as experimental data, it is possible to determine δ by an iterative calculation using eqs. (37) and (39) or eqs. (36) and (38), respectively, together with Figure 7. The apparent molecular weight M_{APP} and the apparent mean-square radius of gyration $\langle s^2 \rangle_{APP}$ determined from $R_{Uv}(u)$ or $R_{Vv}(u)$ can then be corrected to M_w and $\langle s^2 \rangle_{LS}$, respectively:

$$\lim_{\substack{c=0 \\ \theta=0}} \frac{Kc}{R_{Uv}(u)} = \frac{1}{M_{APP,U}} = \frac{1}{M_w \left(1 + \frac{7}{5}\delta^2\right)} \tag{40}$$

$$\lim_{c=0} \frac{Kc}{R_{Uv}(u)} = \frac{1}{M_{APP,U}} \left(1 + \frac{1}{3} \langle s^2 \rangle_{APP,U} h^2 + \cdots \right) \tag{41}$$

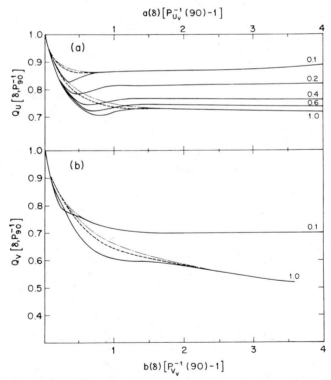

FIG. 7. Depolarization functions Q_V and Q_U, defined in the text, for anisotropic polydisperse rods for polydispersity factors h equal to ∞, —; 4, - - -; and 2, — · —. The value of δ^2 is shown in the figure.

$$\langle s^2 \rangle_{LS} = \frac{1 + \dfrac{7}{5}\delta^2}{1 - \dfrac{4}{5}\delta + \dfrac{47}{35}\delta^2} \langle s^2 \rangle_{APP,U} \tag{42}$$

and

$$\lim_{\substack{c=0 \\ \theta=0}} \frac{Kc}{R_{Vv}(u)} = \frac{1}{M_{APP,V}} = \frac{1}{M_w \left(1 + \dfrac{4}{5}\delta^2\right)} \tag{43}$$

$$\lim_{c=0} \frac{Kc}{R_{Vv}(u)} = \frac{1}{M_{APP,V}} \left(1 + \frac{1}{3}\langle s^2 \rangle_{APP,V} h^2 + \cdots\right) \tag{44}$$

$$\langle s^2 \rangle_{LS} = \frac{1 + \dfrac{4}{5}\delta^2}{1 - \dfrac{4}{5}\delta + \dfrac{4}{7}\delta^2} \langle s^2 \rangle_{APP,V} \tag{45}$$

We remark that the Cabannes factor defined as

$$C_u(\theta) = \frac{6 + 6\rho_u(\theta)}{6 - 7\rho_u(\theta)} \tag{46}$$

with $\rho_u(\theta) = R_{Hu}(\theta)/R_{Vu}(\theta)$ has sometimes been used to correct the scattering $R_{Uu} = R_{Vv} + R_{Vh} + R_{Hv} + R_{Hh}$ with natural light when treating light-scattering data on anisotropic polymers according to the relation

$$\lim_{\substack{c=0 \\ \theta=0}} \frac{Kc}{(1 + \cos^2\theta)R_{Uu}(\theta)} = \frac{1}{M_{APP}} = \frac{1}{M_w \left(1 + \dfrac{13}{5}\delta^2\right)} \tag{47}$$

with

$$1 + \frac{13}{5}\delta^2 = C_u^{-1}(\theta) \tag{48}$$

Although this correction is proper when $C_u(\theta)$ extrapolated to zero scattering angle is used, it may lead to serious error in application to anisotropic macromolecules if the observable Cabanne factor at 90° is used instead, which is the usual practice. Since the reciprocity relation $R_{Hv} = R_{Vh}$ gives $\rho_u = \rho_U$, the procedure given above should be used to estimate δ from ρ_u or ρ_U for calculation of M_w from M_{APP}.

EXPERIMENTAL

Materials

A sample of BBL supplied by Dr. R. L. Van Deusen, of Wright-Patterson Air Force Base, Ohio, has been used throughout. A portion of this polymer was first dissolved in the "leuco reagent" [18] (polymer concentration 0.5 g/dl) and held for 4 days at room temperature to cleave any residual imide bonds resulting from incomplete cyclization. The treated polymer was then precipitated, washed until neutral, and dried.

Methane sulfonate acid (MSA) used to prepare solutions of BBL and for chromatography was distilled under vacuum (about 0.005 mm Hg) and stored under dry nitrogen until used.

Fractions of BBL were prepared by exclusion chromatography from solution (1.5 g/dl) in methane sulfonic acid using an apparatus described previously [2]. About 20 fractions 17.5 ml in volume each were collected for each injection (20 ml of solution) on the column. Five separate preparations are used in this study. The untreated polymer was used to prepare three of these, and the polymer treated by the leuco reagent was used to prepare the other two series, which are designated by the prefix L. For two of the five preparations, including one preparation each of the polymer as received and after treatment with the leuco reagent, the eluent was used as obtained from the chromatograph—although the polymer concentration of such solutions was low, it was high enough for satisfactory light scattering and viscometry. Fractions of these preparations are designated by the codes 533 and L-534/36.

With three of the preparations of the eluent from the column was recovered by precipitation in methanol, washed with water until neutral, and dried. These preparations are indicated by the suffix D. With series 532-D the precipitation in methanol was accomplished rapidly by adding the eluent rapidly to a large excess of methanol, in which case it precipitated as a blue mass. Fractions of series L-534/36-D were recovered by slow addition of methanol to the eluent from the column, bringing it to a point where the polymer formed a red precipitate that was allowed to settle over a period of several days. The red preciptate was then placed in water, at which point it became blue. It was then washed with water until neutral and dried.

Fractions of series S-500-D are secondary fractions prepared with the polymer as received. Primary fractions of several injections were collected and combined according to elution volume. They were precipitated by slow addition of methanol to the pont where the red precipitate was formed. The latter was collected on a frit and redissolved in MSA. Residual methanol was removed under vacuum at room temperature. It was necessary to add about 50% by volume of benzene sulfonc acid to effect dissolution of the primary fractions with elution volume greater than about 50. The solutions of the primary fractions were reinjected on the chromatograph, and the eluent was collected and combined according to elution volume. The polymer was precipitated and recovered by the procedure used with series 532-D.

Light Scattering and Fluorescence

Light-scattering measurements were made using procedures and instrumentation described elsewhere [2, 19] for studies on colored, fluorescent polymers. Small, centrifugable scattering cells were used. The scattering with 6328-Å wavelength incident light determined over the angular range 25–135° was corrected for absorption and fluorescence by procedures discussed in detail previously [2, 20]. Concentrations, which are less than 5×10^{-4} g/ml in all cases, were determined by optical density measurements using a Cary model 14 recording spectrophotometer. The "excess" scattering functions $R_{Vv}(\theta,c)$ and $R_{Vh}(\theta,c)$ were calculated using the scattered intensity from the solution, corrected for absorption and fluorescence, less that from the solvent. Although this procedure ignores certain corrections due to polymer-solvent orientation and deviations of the local field in the solution from that given by the Lorentz-Lorenz approximation, it is believed that this neglect has little effect on the estimate for the dependence of δ on molecular weight [21], which is our primary concern with respect to the depolarized light scattering. Plots of $Kc/R_{Uv}(\theta,c)$ versus $\sin^2 \theta/2$ were extrapolated to zero angle to give intercepts $Kc/R_{Uv}(0,c)$. These intercepts were then plotted against the concentration to obtain the intercepts $[Kc/R_{Vv}(0,c)]^0 = M_{APP}^{-1}$ as the intercept at infinite dilution. The initial slopes of the plots of $Kc/R_{Uv}(\theta,c)$ versus $\sin^2 \theta/2$ were used to determine $\langle s^2 \rangle_{APP}$ in the standard way.

These extrapolations for the higher-molecular-weight polymers were checked against extrapolations of $Kc/R_{Uv}(\theta,c)]^2$ versus $\sin^2 \theta/2$ suggested by the analysis given above. The agreement between values of M_{APP} and $\langle s^2 \rangle_{APP}$ determined in these two ways was satisfactory.

The refractive index increment dn/dc determined with a differential refractometer [20] with 6328-Å wavelength light was found to be 0.82 ml/g at 25°.

The light-scattering photometer was used to determine the depolarization of the fluorescence from solutions of BBL. The fluorescence spectra for excitation at several wavelengths was determined with an Aminco-Bowman fluorometer. For depolarization studies the fluorescence was excited with 6328-Å wavelength incident light, and an interfeerence filter with a sharp band rejection for this wavelength light was used to eliminate scattered light and permit direct observation of the fluorescence. The intensities $I_\parallel$ and I_+ between parallel and crossed polars were determined and used to calculate the dimensionless fluorescence emission anisotropy r defined as [20]

$$r = \frac{I_\parallel - I_+}{I_\parallel + 2I_+} \tag{49}$$

Intrinsic Viscosity

Intrinsic viscosities were determined with suspended level Cannon-Ubbelohde capillary viscometers. With the dried polymers, concentrations could be made sufficiently high to permit extrapolations of η_{sp}/c and $\ln \eta_{rel}/c$ versus c to infinite dilution to determine the intrinsic viscosity $[\eta]$ according to the usual relations:

$$\frac{\eta_{sp}}{c} = [\eta] + k'[\eta]^2 c + k''[\eta]^3 c^2 + \cdots \tag{50}$$

$$\frac{\ln \eta_{rel}}{c} = [\eta] - \left(\frac{1}{2} - k'\right)[\eta]^2 c + \left(\frac{1}{3} - k' + k''\right)[\eta]^3 c^2 + \cdots \tag{51}$$

Here η_{rel} is the relative viscosity, and $\eta_{sp} = \eta_{rel} - 1$. These simultaneous plots were required to yield the same value of k' to within 0.005. Values of k' were in the range 0.34–0.4. With the samples used directly from the chromatographic column, the concentration was usually too low to permit dilution to be made ($\eta_{rel} \approx 1.1$). To treat these data, eqs. (50) and (51) were rearranged by subtracting eq. (51) from eq. (50) and taking the square root of the result to give

$$\left\{\frac{2(\eta_{sp} - \ln \eta_{rel})}{c^2}\right\}^{1/2} = [\eta] - \left(\frac{1}{3} - k'\right)[\eta]^2 c + \cdots \tag{52}$$

in which terms of order c are minimized if k' is near $\frac{1}{3}$. Equation (52) was used to compute $[\eta]$ by a "one-point" method in which terms $O(c)$ on the rhs were neglected.

RESULTS

Values of M_{APP}, $\langle s^2 \rangle_{APP}$, and δ^2 determined by light scattering are listed in Table I, along with the corrected parameters M_w and $\langle s^2 \rangle_{LS}$ calculated with eqs. (39) and (41), respectively. The anisotropy δ was determined from $\rho_U(90°)$ using eq. (37) and Figure 7(a) by an iteration in which δ was first estimated with eq. (37) by putting $Q_U = 1$. This was used to obtain an estimate of Q_U from Figure 7(a), which was used to calculate a new value of δ with eq. (37). The process was continued until δ did not change on successive iterations. It may be noted that δ is zero for the series S-500-D fractions, in contrast to the data on the fractions of the other preparations. Moreover, the values of M_w for this polymer do not decrease monotonically with increasing elution volume V_e of the fractions, as would be expected and is observed with the other fractions.

Plots of $\log [\eta]$ and $\log M_w$ versus V_e are given in Figures 8 and 9, respectively. A good correlation is observed between $\log [\eta]$ and V_e for the data on fractions of series 533, 532-D, and S-500-D, in Figure 8, with the data on the latter showing the greatest deviation for the average. Similarly, a satisfactory correlation is observed between $\log M_w$ and V_e for the data on fractions of series 533, 532-D, and L-534/36 in Figure 9. By contrast, however, data on fractions of series L-534/36-D and S-500-D deviate markedly from this correlation between M_w and V_e and are themselves scattered. Assuming that this deviation is caused

TABLE I
Dilute Solution Parameters for Solutions of BBL in Methane Sulfonic Acid

Polymer	Elution volume on preparative scale column	$10^{-4} M_{APP}$	$10^{12} \langle s^2 \rangle_{APP}$ (cm^2)	δ^2	$10^4 M_w$	$10^{12} \langle s^2 \rangle_{LS}$ (cm^2)
533	40	5.3	90.4	0.12	4.5	119
	45	2.7	18.4	0.24	2.0	26.5
	53	1.8	6.2	0.39	1.16	9.4
	65	1.2	—	0.68	0.61_7	—
532-D	35–37	13.4	236	0.20	10.5	330.1
	44–45	2.8	16.2	0.24	2.1	23.3
	52–53	2.07	7.1	0.35	1.4	10.7
	64–67	1.25	~3.9	0.54	0.71	~6
L-534/36	40	4.8	39.1	0.40	3.1	59.4
	45	3.0	15.5	0.46	1.8	23.7
	50	2.0_7	10.9	0.46	1.2_6	16.7
	60	1.9	—	0.90	0.84	—
	70	1.4	—	1.0	0.58	—
L-534/36-D	40	12.0	133	0.31	8.4	197
	50					
	60	10.2	181	0.42	6.5	275
S-500-D	35–36	326	428	0		
	41–52	28.3	158	0		
	43–44	18.6	130	0		
	45–46	35.7	220	0		
	49–50	25.7	970	0		
	51–52	26.3	1560	0		
	53–54	11.4	445	0		
Whole	—	8.0	103	0.38	5.2	160

by association, an estimate of the degree of association ν can be obtained by comparison of M_w with the molecular weight M_e corresponding to the elution volume V_e according to the correlation obtained with fractions of the other four preparations. This calculation gives $\nu = M_e/M_w$ ranging from 2 to 20, with the exception of the first fraction of series S-500-D, for which $\nu = 410$.

We believe that the slow coagulation step in common to series L-534/36-D and S-500-D fractions is the source of the association encountered with these fractions. The red form of the precipitated polymer obtained in the slow coagulation step is presumably a form in which the polymer is highly protonated, as it certainly is in solution [4]. Apparently, the molecular mobility in this state is sufficient to permit the formation of rather stable aggregates, as contrasted to the situation when coagulation is rapid. This is consistent with the observation that the primary fractions of series 500-D were insoluble in MSA but soluble in benzene sulfonic acid.

The fluorescence and absorption spectra of a solution of BBL in MSA (0.03 g/liter) are shown in Figure 10. The emission spectra are uncorrected for absorption. It may be observed that the maximum emission occurs at 476 nm, irrespective of the wavelength of the excitation light over the range 240–550 nm

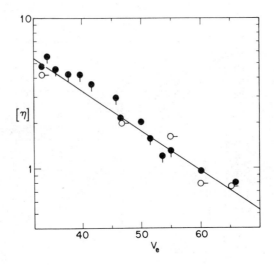

FIG. 8. Intrinsic viscosity versus elution volume for solutions of BBL fractions in MSA: samples 532-D, ●; 533, O-; S-500-D, ◐.

studied here. The reciprocal emission anisotropy r^{-1} is given as a function of $\tau RT/\eta_0[\eta]M_w$ in Figure 11 for two solutions of BBL fractions in MSA with $[\eta]M_w$ equal to 1.0×10^5 and 2.3×10^4, respectively. Data were taken over the temperature range 15–86°. Here η_0 is the solvent viscosity at temperature T, and τ is the lifetime of the excited state. The latter was determined to be 0.9 nsec by a pulse-decay experiment in which the decay of the fluorescence following excitation by a pulse was fitted to the exponential relation

$$I(t) = I(0) \exp\left(\frac{-t}{\tau}\right) \tag{53}$$

using instrumentation and a deconvolution technique discussed elsewhere [22].

DISCUSSION

BBL Conformation from Light-Scattering Data

According to Sicree et al. [23], inspection of molecular models of BBL suggests that the chain may be nearly represented by a two-dimensional model in the sense that the entire molecule may be nearly planar, with elements with the structure

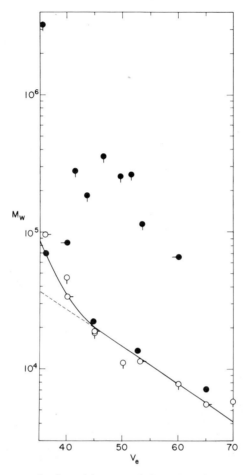

FIG. 9. Weight-average molecular weight versus elution volume for solutions of BBL fractions in MSA. Points identified in the caption to Figure 8 except for samples L-534/36, ◊, and L-534/36-D, —●.

or its isomer formed by the mirror image of the structure shown, randomly connected head to head and tail to tail to form the chain. The chain can then be modeled by a succession of segments of length l (the dotted line in the structure above) and mass 167, all segments lying in a plane, and with angle $\pi - \alpha$ between successive segments. Of course, α is not a constant but depends on the isomers joined as neighbors. Sicree et al. conclude that α should lie in the range 5°–10°, making the BBL chain nearly rodlike as well as rigid.

The mean-square radius of gyration of an ensemble of such a rigid chain containing all possible structural isomers with average direction cosine $\langle \cos \alpha \rangle$ is given by eq. (5), calculated for a freely rotating chain with average direction cosine $\langle \cos \alpha \rangle$. This result obtains because the average cosine $\langle \cos \varphi \rangle$ of the angle of rotation φ about bonds away from the planar zigzag conformation (with $\varphi = 0$) is zero both for a chain with free rotation and for a chain with rotation re-

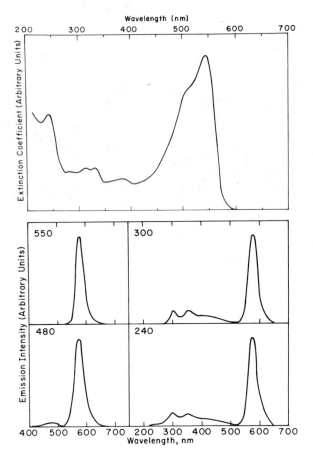

FIG. 10. Upper, absorption spectrum for BBL in MSA (0.03 g/liter); lower, emission spectrum for BBL in MSA (0.03 g/liter) for several excitation wavelengths.

stricted to isomers with $\varphi = 0$ or π, the latter characterizing a planar molecule. This circumstance permits an estimate for $\langle \cos \alpha \rangle$ by comparison of experimental data on $\langle s^2 \rangle / N$ versus N (here $N = M/167$) with the calculated curve for $S(N,a)$ versus $N(1 - a)/a$ given in Figure 2. This is conveniently accomplished graphically by superposing a transparency of the experimental data of $\langle s^2 \rangle / N$ versus N on Figure 2 and noting the values of $\langle s^2 \rangle / N$ and N that correspond to $S(N,a)$ and $N(1 - a)/a$ equal to unity, respectively, when the two graphs superpose. These give $(\langle s^2 \rangle / N) = l^2 (1 + a)/6(1 - a)$ and $a/(1 - a)$, respectively.

Use of eq. (3) to analyze data on δ as a function of N for the rigid planar chain is not strictly justified, since $\langle \cos^2 \phi \rangle$, which enters into the calculation for δ, is equal to unity for the two-dimensional chain as compared with one-half for the three-dimensional freely rotating chain used to compute eq. (3). Nonetheless, since we are interested in δ only for small values of $N(1 - a)/a$ and for a nearly rodlike polymer, we use eq. (3) rather than the considerably more complicated

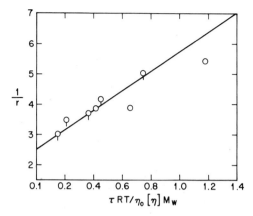

FIG. 11. Reciprocal emission anisotropy r^{-1} versus $\tau RT/\eta_0[\eta]M$ for L534/36; 40, O; L-534/36; 50, O, in MSA.

form [9] valid for all $N(1 - a)/a$ for a chain with hindered rotation with $\langle \cos \phi \rangle = 0$ and $\langle \cos^2 \phi \rangle = 1$. Thus, comparison of experimental data with eq. (3) provides a means to determine δ_0 and $a = \langle \cos \alpha \rangle$. A graphical method is again convenient in which a transparency of the experimental data of δ^2 versus N is superposed on Figure 1 and the values of δ^2 and N corresponding to δ^2/δ_0^2 and $N(1 - a)/a$ equal to unity, respectively, are noted. These give δ_0^2 and $a/(1 - a)$, respectively.

The analysis of data on BBL is complicated by the effects of unknown molecular-weight heterogeneity in the fractions. Assuming that the molecules are essentially rodlike over the range of molecular weights to be analyzed, the appropriate N_{AVG} to be used in the analyses of $\langle s^2 \rangle_{LS}$ and δ described above is $\sqrt{N_z N_{z+1}}$. Thus, the intersection point where $(\delta/\delta_0)^2$ and $N_{AVG}(1 - a)/a$ are unity shown in Figure 12 gives $\delta_0 = 1$ and $am_0/(1 - a) = 15,850$

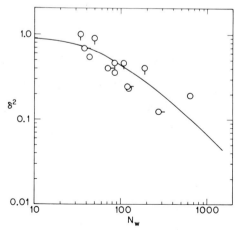

FIG. 12. Anisotropy δ^2 versus weight average degree of polymerization for fractions of BBL. Points, are identified in the caption to Figure 9. The curve represents eq. (3).

$\sqrt{M_z M_{z+1}/M_w^2}$, respectively. Similarly, the intersection where $S(N,a)$ and $N_{AVG}(1 - a)/a$ are unity shown in Figure 13 gives $(\langle s^2 \rangle_{LS}/M_w)_\infty = 49 \times 10^{-16}$ cm²/dalton, or $am_0/(1 - a) = 103,500/\sqrt{M_z M_{z+1}/M_w^2}$, with $l = 6.3$ Å and $2a/(1 + a) \approx 1$, and $am_0/(1 - a) = 15,850 \sqrt{M_z M_{z+1}/M_w^2}$, respectively. In these fits, the data in Figures 12 and 13 were required to give the same values for $N_{AVG}(1 - a)/a$ and $\delta \leq 1$. Solving for a and $M_z M_{z+1}/M_w^2$, we find 6.5 for the latter and $am_0/(1 - a) = 40,500$, or $\alpha = 5.19°$.

The value of α is in good accord with the prediction of Arnold et al. that α should lie in the range 5–10°. Similarly, the value $\delta_0 = 1$ appears to be reasonable for BBL. The fractions appear to have a long high-molecular-weight tail. For example, $M_z M_{z+1}/M_w^2$ is too large to be fitted with a Zimm-Schulz molecular-weight distribution. Since the functions δ and S shown in Figures 1 and 2 can also be calculated with the wormlike model, the data can be interpreted just as well in terms of the persistence length ρ. Thus, the correspondence $nb/\rho = M(1 - a)/am_0$ noted above can be used to estimate the persistence length ρ from $am_0/(1 - a) = 40,500$ with the relation

$$\rho M_L = \frac{am_0}{(1 - a)} \tag{52}$$

where

$$M_L = \frac{M}{L}$$

is the mass per unit contour length $L = nb$ and m_0 is the mass per repeat unit ($M = Nm_0$). Putting $M_L = m_0/l = 26.51$ daltons/Å, which is equivalent to the use of the Kuhn approximation [14c] $nb = Nl$, we find a persistence length of 1530 Å. Since ρ is long compared with Nl, which lies in the range 200–4000 Å for the polymers studied, the use of rodlike chain statistics appears to be reasonable.

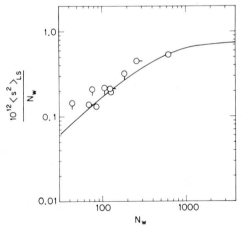

FIG. 13. The ratio $\langle s^2 \rangle_{LS}/N_w$ versus N_w for fractions of BBL. Points are identified in the caption to Figure 9. The curve represents eq. (5).

We have not included the data on fractions of series S-500-D in the analysis above, since these appear anomalous in three respects: (1) neither M_w or $\langle s^2 \rangle_{LS}$ are monotone functions of V_e, (2) δ is rather small in comparison with values on fractions of the other polymers at comparable V_e, and (3) the ratio $\langle s^2 \rangle_{LS}/N_w$ does not correlate well with the data obtained with the other fractions. These effects can all be attributed to association. It appears that the steps used to isolate and recover the primary and secondary fractions in this series or the narrower molecular-weight distribution of the secondary fractions led to some irreversible association. The low values of δ suggest that the aggregates are not formed with the chains packed in a parallel array, but rather more as in a brush heap. At least some of the aggregates are impervious to dissolution in known solvents for BBL.

Intrinsic Viscosity

The molecular-weight dependence of the intrinsic viscosity of the BBL fraction in MSA is shown in Figure 14. Since $[\eta]$ was not always measured on the fraction used for light scattering, values of $[\eta]$ have been interpolated using the correlation in Figure 8. The Mark-Houwink exponent $\partial \ln [\eta] / \partial \ln M$ decreases with increasing M, exhibiting a value greater than unity at low M. The data for fractions of L-534/36D and S-500-D tend to scatter and fall below the correlation for the other fractions, owing to the effects of the association present in solutions of these polymers. Since the molecular conformations are essentially rodlike, the relation for $[\eta]$ calculated with the wormlike model can be used to analyze the data. There are several variations of such calculations, carried out at different

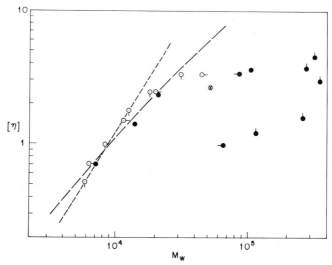

FIG. 14. Intrinsic viscosity versus molecular weight for fractions of BBL in MSA. Points are identified in Figure 9 except for the data on the unfractionated polymer, ⊗. Curves calculated according to Yamakawa and Fujii [28] with $d = 12$ Å, $d/2\rho = 0.03$, $- -$, and $d = 2.2$ Å, $d/2\rho = 0.001$, $----$.

levels of rigor and based on different approximations [14b, 24–28]. For chains with a rodlike or wormlike conformation, the theories are conveniently expressed in terms of a contour length L, the mass M_L per unit contour length, and one or more additional parameters such as the hydrodynamic diameter d, the persistence length ρ, and the equivalent segment length b. For example, $[\eta]$ (in dl/g) can be written in the form

$$[\eta] = K_R \left(\frac{L}{d}\right)^2 f\left(\frac{L}{d}, \frac{\rho}{d}, \frac{b}{d}\right) \tag{53}$$

for several models, where $K_R = \pi N_A d^2/100 M_L$ and f may be a function of the enumerated variables. Thus

$$f = \frac{2}{45} \frac{1}{\ln (L/d)} \tag{54}$$

for long, rigid rods, with $b = d$,

$$f = \frac{2}{45} \frac{0.25}{\ln (L/d) - 2.72 + 0.66(b/d) - \ln (b/d)}$$
$$+ \frac{0.75}{\ln (L/d) - 2.72 + 1.33(b/d) - \ln (b/d)} \tag{55}$$

for wormlike chains in the limiting case $L/\rho \ll 1$ [24], and

$$f = \frac{(3.998/24) S(L/\rho)}{(L/\rho)[(b/d) + 0.3897 \sqrt{L/\rho\, \phi}]} \tag{56}$$

for wormlike chains with larger L/ρ [25], where ϕ is a function of L/ρ and $\lambda = b/2\rho$ (see Appendix) and $S(L/\rho)$ is given by eq. (6b). In addition, Yamakawa and Fujii [28] have calculated $[\eta]$ for a wormlike cylinder model with results that can be cast in the form of eq. (53) with f given numerically or, for $L/2\rho < 2.28$, by

$$f = \frac{h_1(L/\rho)}{h_2(L/d)} \tag{57}$$

with

$$h_1(x) = x^{-4}\left[\exp(-x) - 1 + x - \left(\frac{x^2}{2}\right) + \left(\frac{x^3}{6}\right)\right] \tag{58}$$

$$h_2\left(\frac{L}{d}\right) = 1 - \frac{1.839}{\ln (L/d)} + \frac{8.241}{\ln^2 (L/d)} - \frac{32.863}{\ln^3 (L/d)} + \frac{41.1}{\ln^4 (L/d)} \tag{59}$$

where $x = L/2\rho$. Equations (56) and (57) are similar if $b = d$ in the former.

Comparison of experimental data on $[\eta]$ versus M with the theory is hampered by the need to know several parameters, including M_L, d; ρ, and, with some models, b. To make the problem tractable, we will deal only with models where b does not enter or otherwise set $b = d$. It is convenient to express the parameters in eq. (53) as $K_R = \pi N_A (M_L d)^2/100 M_L^3$ and $L/d = M/M_L d$. Then, given an estimate for M_L, data on $[\eta]$ versus $M_\eta = \sqrt{M_z M_w}$ can be compared with the theoretical curves of $(M_L d)^{-2}[\eta]$ versus $(M_L d)^{-1}M$ to determine ρ and d (here, M_η is estimated with the value $M_z M_{z+1}/M_w^2 = 6.5$ given above). For example, the curves shown in Figure 14 were calculated with the numerical values of f given by Yamakawa and Fujii with the estimate $M_L = m_0/l$, using two sets of the parameters ρ and d: $\rho = 196$ Å and $d = 12$ Å for one curve and $\rho = 1065$ Å and $d = 2.2$ Å for the other. Similar estimates are obtained with eq. (58), with ρ being a little smaller for a given estimate for d than the latter. None of the theoretical curves match the severe downward curvature at high M exhibited by the data on the "dried specimens," but this is not unexpected, since these are believed to be aggregated, giving erroneously large M_w by light scattering. The value of $\rho = 1065$ Å is in reasonable accord with the estimate $\rho = 1530$ Å obtained from the light-scattering data, but the corresponding value of $d = 2.2$ Å is rather small unless some special form of interaction between the polymer and the acidic solvent is assumed. The significance of this finding is not certain, however, since the estimate of d is sensitive to the choice of M_L, for a given ρ. For example, if M_L is increased from 26.5 dalton/Å, corresponding to $m_0 = 167$ and $l = 6.3$ Å, to 34.3 dalton/Å, then $d = 17$ Å when $\rho = 1710$ Å. The larger value of M_L could be interpreted, for example, as reflecting one associated solvent molecule per two repeating units.

Fluorescence Anisotropy

The fluorescence emission anisotropy r for a rigid molecule depends on the intrinsic anisotropy r_0, the lifetime τ of the excited state, and the rotatory diffusion constant D_R according to the relation [20]

$$r = r_0[1 + 6D_R\tau]^{-1} \tag{60}$$

The intrinsic anisotropy r_0, which depends on the angle β between the absorption and emission vectors, is given by

$$r_0 = 0.2[3\cos^2 \beta - 1] \tag{61}$$

and for rods [14b]

$$D_R = \frac{6}{4500} \frac{RT}{\eta_0[\eta]M} \tag{62}$$

Combination of eqs. (60) and (62) gives the relation

$$\frac{1}{r} = \frac{1}{r_0} + \frac{36}{4500r_0} \frac{\tau RT}{\eta_0[\eta]M} \tag{63}$$

which motivated the plot of the experimental data given in Figure 11. In principle, eq. (63) provides a means to determine the molecular weight of a rodlike polymer given r as a function of $\tau RT/\eta_0[\eta]$.

Although the experimental data in Figure 11 extrapolate to a reasonable value of $r_0 = 0.39$, giving β nearly zero, as might be expected for a rodlike polymer, the experiment initial tangent $\partial(r^{-1})/\partial(\tau RT/\eta_0[\eta]M_w)$ is a hundredfold larger than the value $36/4500r_0 = 0.02$ expected according to eq. (63), with $r_0 = 0.4$.

Taken at its face value, this descrepancy would imply that the mobility of BBL is far greater than would be expected for a rodlike molecule. It is necessary, however, to consider the effects of molecular-weight distribution. An average emission anisotropy $\bar{r}$ can be calculated as

$$\bar{r} = r_0 M_n \int_0^\infty \frac{M^{-1}f(M)}{1 + 6\tau D_R(M)} dM \qquad (64)$$

with $D_R(M)$ given by eq. (62). Assuming that $[\eta] \propto M^2$ and $\tau D_R(M)$ is small compared with unity, so that $[1 + 6\tau D_R(M)]^{-1} \simeq 1 - 6\tau D_R(M)$, eq. (64) can be put in the form

$$\frac{\bar{r}}{r_0} \cong 1 - \frac{36}{4500} \frac{\tau RT}{\eta_0 [\eta] M_R} \qquad (65a)$$

with

$$M_R^{-1} \equiv M_n M_w M_z \int_0^\infty M^{-4}f(M) \, dM \qquad (65b)$$

revealing the predominance of the low-molecular-weight species in determining $\bar{r}$. Indeed, if eq. (17) is used for $f(M) \, dM$, the integral in eq. (65) must be calculated with a lower bound $M_1 > 0$ if $h \leq 3$ in order to maintain comvergence. In general, with eq. (17),

$$M_R^{-1} = \frac{1}{M_n} \left(\frac{M_z}{M_w}\right) \left(\frac{M_w}{M_n}\right)^2 \{h^3 \Gamma[(h-3), (h+1)]M_1/M_w/\Gamma(h)\}$$

where $\Gamma(a,b)$ is the incomplete gamma function. If $h > 3$, then M_1 can be put equal to zero, and the factor in curly brackets is equal to $h^3/(h-1)(h-2)(h-3)$. Since this calculation is very sensitive to the form used for $f(M)$ at low M, we cannot reliably estimate the integral in eq. (65) with an assumed $f(M)$. Thus, we conclude that data on the emission anisotropy $\bar{r}$ for a polydisperse rodlike polymer as a function of $\tau RT/\eta_0[\eta]$ cannot be analyzed to give one of the standard molecular-weight averages, e.g., M_n, M_w. The average molecular weight M_R that can be obtained is heavily weighted toward low-molecular-weight components in the distribution.

APPENDIX

The function

$$\varphi(n,\lambda') = \frac{15(\pi/3)^{1/2}}{4(3 - \sqrt{2})} \cdot \frac{1}{n^2 \sqrt{\lambda' n}}$$

$$\times \left\{ \sum_{k=1}^{n-1} (k^2 + k - nk + 2n) \frac{\Psi(x)}{\sqrt{\gamma(x)}} + \sum_{k=1}^{n-1} \left[\left(\frac{n^2}{2}\right) - 2k^2 + n\right] \frac{\Psi(x)}{\sqrt{\gamma(x)}} \right\}$$

where

$$\Psi(x) = \frac{0.4270 + 0.0212\{45x^2 - 156x + 214 - 54(4 + x)e^{-x} + 2e^{-3x}\}}{\gamma^2(x)}$$

$$\gamma(x) = x - 1 + e^{-x}$$

TABLE II

$n\backslash\lambda'$	1	5	10	15	20	30	50	100
10	−0.1413	+0.1847	+0.3488	+0.4603	+0.5498	+0.5951	+0.9191	+1.3223
8	−0.3297	−0.1721	−0.1511	−0.1490	−0.1534	−0.1644	−0.1886	−0.2420
6	−0.6324	−0.7769	−0.9992	−1.1853	−1.3470	−1.6237	−2.0697	−2.8993
4	−1.2115	−1.9866	−2.6980	−3.2605	−3.7400	−4.5503	−5.8436	−8.2317

and $x = k/\lambda'$ has been tabulated in reference R for $n \geq 10$ for several values of $\lambda' = (2\lambda)^{-1}$ in the range $1 \leq \lambda' \leq 100$. Additional values calculated for $n = 4, 6, 8$, and 10 are entered in Table II. In addition, for $L/\rho > 6$, the empirical relation

$$\sqrt{\frac{L}{\rho}} = \phi = \sqrt{\frac{L}{\rho}} - 3.37(1 + 0.451 \ln \lambda)$$

can be used [29].

The assistance of S. M. Liwak on the light-scattering and viscosity measurements is gratefully acknowledged. In addition, we are indebted to Dr. L. K. Patterson, then of the Bushy Run Radiation Laboratory, Carnegie-Mellon University, for measurement of the fluorescence lifetime. This study was supported in part by a contract from the Air Force Materials Laboratory, Wright-Patterson Air Force Base, Ohio, and in part by a grant from the National Science Foundation.

REFERENCES

[1] R. L. Van Deusen, *J. Polym. Sci. Part B, 4*, 211 (1966).

[2] G. C. Berry, *Discuss. Faraday Soc., 49*, 121 (1970).

[3] C. P. Wong and G. C. Berry, *Prepr. Div. Org. Coat. Plast. Am. Chem. Soc., 33*(1), 215 (1973).

[4] G. C. Berry and P. R. Eisaman, *J. Polym. Sci. Polym. Phys. Ed., 12*, 2253 (1974).

[5] G. C. Berry, *J. Polym. Sci. Polym. Phys. Ed., 14*, 451 (1976).

[6] R. L. Van Deusen, A. K. Gorns, and A. J. Surei, *J. Polym. Sci. Part A-1, 6*, 1777 (1968).

[7] P. Horn, *Ann. Phys., 10*, 386 (1955).

[8] H. Utiyama and M. Kurata, *Bull. Inst. Chem. Res. Kyoto Univ., 42*, 128 (1964).

[9] M. V. Volkenstein, "Configurational Statistics of Polymer Chains," Wiley-Interscience, New York, 1965, p. 443.

[10] P. J. Flory, "Statistical Mechanics of Chain Molecules," Wiley-Interscience, New York, 1969, p. 353.

[11] H. Benoit, *C. R. Acad. Sci., 236*, 687 (1953).

[12] K. Nagai, *Polym. J., 3*, 67 (1972).

[13] H. Benoit and P. Doty, *J. Phys. Chem., 57*, 958 (1953).

[14] H. Yamakawa, "Theory of Polymer Solutions," Harper & Row, New York, 1971 (a) p. 56; (b) Chap. 6; (c) p. 46.

[15] H. Benoit, *Makromol. Chem., 18–19*, 397 (1956).

[16] G. V. Schulz, *Z. Phys. Chem. Abt. B, 43*, 25 (1939).

[17] M. Goldstein, *J. Chem. Phys., 21*, 1255 (1953).

[18] G. C. Berry and S.-P. Yen in "Symposium on Polymerization and Polycondensation Processes (Adv. Chem. Ser., No. 91) R. F. Gould, Ed., American Chemical Society, Washington, D.C., 1970, p. 734.

[19] E. F. Casassa and G. C. Berry in "Polymer Molecular Weights," Part I, R. E. Slade, Ed., Dekker, New York, 1975, p. 161.

[20] D. J. R. Laurence in "Physical Methods in Macromolecular Chemistry," B. Carroll, Ed., Dekker, New York, 1969, Chap. 5.

[21] I. Fortelny, *J. Polym. Sci. Polym. Phys. Ed., 14*, 1931 (1976).

[22] L. K. Patterson and E. Vieil, *J. Phys. Chem., 77*, 1191 (1973).

[23] A. J. Sicree, F. E. Arnold, and R. L. Van Deusen, *J. Polym. Sci. Polym. Chem. Ed.*, *12*, 265 (1974).

[24] J. E. Hearst, *J. Chem. Phys.*, *40*, 1506 (1964).

[25] Y. E. Eizner and O. B. Ptitsyn, *Vysokomol. Soedin.*, *4*, 1725 (1962).

[26] J. G. Kirkwood and P. L. Auer, *J. Chem. Phys.*, *19*, 281 (1951).

[27] S. F. Kurath, C. A. Schmitt, and J. J. Backhuber, *J. Polym. Sci. Part A, 3*, 1825 (1965).

[28] H. Yamakawa and M. Fujii, *Macromolecules, 7*, 128 (1974).

[29] D. W. Tanner and G. C. Berry, *J. Polym. Sci. Polym. Phys. Ed., 12*, 941 (1974).

PROPERTIES OF SOME RODLIKE POLYMERS IN SOLUTION

C.-P. WONG, H. OHNUMA,* and G. C. BERRY

Department of Chemistry, Carnegie-Mellon University, Pittsburgh, Pennsylvania 15213

SYNOPSIS

The properties of poly(1,4-phenylene terephthalamide) (PPTA), Kevlar, and a poly(benzobisoxazole) (PBO) in solution are investigated. Molecular characterization using light scattering and viscometry on dilute solutions shows PBO and PPTA to have a rodlike conformation. Both are prone to interchain aggregation with the tendency for association apparently increasing with the ionic strength of the solvent. Rheological experiments are reported on disordered and ordered concentrated solutions of Kevlar and PBO. The dependence of the viscosity and recoverable compliance on shear rate is discussed and compared with that of flexible-chain polymers. The dependence of the viscosity and recoverable compliance of PBO solutions on temperature and concentration is discussed.

INTRODUCTION

Solutions of poly(1,4-phenylene terephthalamide) (PPTA) and the poly-(benzobisoxazole) (PBO) have in common some unusual properties in both dilute and moderately concentrated solutions:

PBO

PPTA

In this report we discuss some of these properties, including light scattering and viscometry with dilute solutions, and rheological characterization of moderately concentrated solutions. It is shown that appreciable association obtains in many solutions, even at low polymer concentration. Data (apparently) free of association in dilute solution are discussed in terms of a wormlike chain model to assess the rodlike character of PPTA and PBO. The rheological properties to be considered include the dependence of the viscosity η_κ and the steady-state recoverable compliance R_κ on the shear rate κ. Here, R_κ is equal to the recovery γ_{R_κ} after

* Permanent address: Department of Physics, Nagoya University, Nagoya, Japan.

Journal of Polymer Science: Polymer Symposium 65, 173–192 (1978)
0360-8905/78/0065-0173$01.00

cessation of steady-state flow, divided by the shear stress $\kappa\eta_\kappa$. Data on both isotropic and anisotropic solutions are discussed.

EXPERIMENTAL

The discussion of experimental procedures is abbreviated here, since detailed discussion of most of these appear elsewhere in this symposium [1, 2]. Polymers were obtained from three sources: (1) PBO polymers were provided by Dr. F. E. Arnold, Polymers Branch, Air Force Materials Laboratory, Wright-Patterson Air Force Base, Ohio; (2) Kevlar fiber was provided by the Textile Fibers Department, E. I. du Pont de Nemours and Company, Experimental Station, Wilmington, Delaware; and (3) samples of PPTA were polymerized in this laboratory using known procedures [3].

The PPTA polymerizations were carried out as follows: A solution of 12.5 mmoles of p-phenylene diamine and 5 g LiCl in 120 ml hexamethylene phosphoramide (HMPA) and 30 ml N-methyl pyrrolidone-2 (NMP) was degassed with nitrogen (HMPA should be treated as a potentially carcinogenic reagent [3]. After addition of 12.5 mmoles terephthaloylchloride in 30 ml NMP at 0°, the stirred mixture was brought to room temperature over a period of about 1 hr and then was held at 100° for 12 hr. The product was precipitated in 500 ml of cold methanol in a blender, collected and washed by filtration, and dried. It was then redissolved in 100 ml sulfuric acid, filtered, reprecipitated with methanol, dialyzed against water, and dried.

The Kevlar fiber was exhaustively washed with chloroform to remove possible surface agents prior to use. The infrared absorption spectra of the Kevlar fiber used was identical with the PPTA prepared in our laboratory, and distinct from poly(p-benzamide) also prepared in our laboratory. Spectra were obtained with polymer powders in KBr pellets.

Methane sulfonic acid, MSA, and chlorosulfonic acid (CSA) were vacuum-distilled and stored under nitrogen prior to use.

The procedures for exclusion chromatography, light scattering, refractometry, and viscometry with dilute solutions in strong acids are described in reference 1, including fluorescence and absorption corrections necessary for the light-scattering analysis and the treatment of light-scattering data for optically anisotropic polymer solutions. The use of salt solutions requires further comment here. Lithium chloride was added to chlorosulfonic acid in the required amount to liberate HCl gas and generate $ClSO_3Li$ in solution. It was found that erratic refractometry data were obtained unless the resultant salt solution was allowed to stand for several days, presumably owing to residual HCl in excess concentration. Consequently, salt solutions were stored under nitrogen in a large vacuum desiccator containing dry silica gel. Refractive index increments dn/dc (25°, 6328 Å wavelength) were found to be 0.474 and 0.254 cm^3/g for PBO and PPTA, respectively, in methane sulfonic acid and 0.502 and 0.287 cm^3/g for PBO and PPTA, respectively, in chlorosulfonic acid. Since addition of LiCl did not cause a perceptible change in dn/dc in chlorosulfonic acid, it was not necessary to treat the salt solution as a mixed solvent in the light-scattering analysis.

Rheological studies were carried out with a cone and plate rheometer [4]. Experimental methods are described elsewhere in this symposium [2] and in the literature [4, 5]. Particular care was taken to protect solutions from moisture during installation in the apparatus by the use of a dry bag attached directly to the rheometer. The instrument was continuously purged with dry argon.

RESULTS AND DISCUSSION

Studies of Dilute Solutions

In the following discussion of the properties of dilute solutions of PPTA and PBO it is convenient to use the familiar wormlike chain model in which the principal characteristic parameters include the chain length L and the persistence length ρ. For example, the use of the wormlike chain model [6, 8] to analyze light-scattering data on rodlike chains is discussed in detail by one of us (GCB) elsewhere [1, 9]. The vertical and horizontal polarized components R_{V_v} and R_{H_v}, respectively, of the Rayleigh ratio with vertically polarized incident radiation are conveniently expressed in terms of the overall chain anisotropy δ and the mean-square radius of gyration $\langle s^2 \rangle$ to terms of order h^2:

$$[R_{V_v}(u,\delta)/KMc]^0 = \left(1 + \frac{4}{5}\delta^2\right) - \frac{1}{3}\left[1 - \frac{4}{5}f_1\delta + \frac{4}{7}(f_2\delta)^2\right]u + o(u^2) \quad (1)$$

$$[R_{H_v}(u,\delta)/KMc]^0 = \frac{3}{5}\delta^2 - \frac{9}{35}(f_3\delta)^2u + o(u^2) \quad (2)$$

where $u = \langle s^2 \rangle h^2$, with $h = 4\pi n/\lambda$. With the wormlike chain model, the parameters δ/δ_0, $\langle s^2 \rangle$, f_1, f_2, and f_3 all depend on the ratio L/ρ. Here δ_0 is the intrinsic anisotropy, it being assumed that the scattering elements have cylindrical symmetry. Since δ decreases rapidly with increasing L/ρ, one can set f_1, f_2, and f_3 equal to the value unity reached for rodlike chains (small L/ρ) for practical purposes in the determination of $\langle s^2 \rangle_{LS}$, M_w, and, in cases of interest here, δ, over the entire range of L/ρ.

Representation of the intrinsic viscosity $[\eta]$ with the wormlike chain model requires the introduction of a chain diameter d and the mass per unit contour length M_L. Then, $[\eta]$ can be put in the form

$$[\eta] = (\pi N_A/100M_L)L^2f(L/d,L/\rho) \quad (3)$$

where $L = M/M_L$. For long rods, in the limit as d/ρ tends to zero, the relation for $f(L/d, L/\rho)$ for the wormlike model can be approximated reasonably well over the range $40 < L/\rho < 10^3$ by $0.0257(d/L)^{0.2}$. In this case, $[\eta]$ is only weakly dependent on d and is proportional to $M^{1.8}$. Consequently, the value of $[\eta]$ should not be solvent-dependent for solutions of rodlike molecules. For larger d/ρ (less rodlike) it is useful to compare bilogarithmic graphs of the experimental data in the form $[\eta]_R = (100M_L^3/N_a\pi)\,[\eta]$ versus M with theoretical graphs of $(L/d)^2f(L/d, L/\rho)$ versus L/d calculated at constant d/ρ [2]. This procedure yields the parameters d and ρ, given M_L. In order to estimate M_L, it is usual to use the Kuhn approximation [6] that the contour length L is equal to lx, where x is the number of chain elements of contour length l and mass m_0. Then, M_L is equal to m_0/l and can be calculated with crystallographic or geometric estimates of l. Alternatively, given accurate data on $\langle s^2 \rangle$ as a function of M, the relation

$$\langle s^2 \rangle = (M^2/12M_L{}^2)S(L/\rho) \tag{4a}$$

$$(y/4)S(y) = 1 - 3y^{-1} + 6y^{-3}[1 - \exp(-y)] \tag{4b}$$

$$\lim_{y=0} S(y) = 1 - \frac{y}{5} + \cdots \tag{4c}$$

can be used together with viscometric estimates of $M_L\rho$ and M_Ld to determine M_L, ρ, and d. Since the latter strategy is not useful in this study, owing to the effects of molecular-weight heterogeneity, the Kuhn approximation is used instead, with M_L equal to 18.45 and 19.15 daltons/Å for PPTA-2 and PBO, respectively.

Data on the fractions of the relatively low-molecular-weight PBO-6 polymer can be used to illustrate the application of these relations. With this polymer, the values of $[\eta]$ in chlorosulfonic acid, methane sulfonic acid, and benzene sulfonic acid were found to be 1.13, 1.0, and 0.9 dl/g, respectively, all reasonably close, in accord with the behavior expected for rodlike chains. Moreover, as shown in Figure 1, the data on $[\eta]$ versus M_w are characterized by a large Mark-Houwink exponent $\partial \ln [\eta]/\partial \ln M_w$ of 1.85, very close to the value expected for a rodlike polymer. Values of about unity for the chain anisotropy δ are also as expected for a rodlike conformation for PBO. Assuming that PBO does adopt a rodlike conformation, the data on $[\eta]$ and the mean-square radius

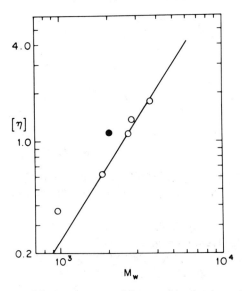

FIG. 1. Intrinsic viscosity $[\eta]$ plotted versus weight-average molecular weight M_w for fractions of PBO-6. The filled circle is for the whole polymer.

TABLE I
Molecular-Weight Parameters for a PBO Polymer

Fraction	X_w	$\dfrac{X_z}{X_w} \simeq \left(\dfrac{X_\eta}{X_w}\right)^2$	$\dfrac{X_{z+1}}{X_z} = \left(\dfrac{X_s}{X_w} \cdot \dfrac{X_w}{X_z}\right)^2$
1 + 2	4.1	10.7	12.9
3	7.8	5.2	20.8
4	11.0	4.7	21.6
5	12.0	5.2	13.5
6	15.8	4.1	12.6
Unfractionated	8.4	8.8	4.9

of gyration $\langle s^2 \rangle_{LS}$ can be used to deduce average degrees of polymerization x_η from eq. (3) and x_s from the relation $\langle s^2 \rangle_{LS} = (x_s l)^2/12$ for comparison with x_w (here, $x = M/m_0$, with $m_0 = 2.34$ and $l = 12.2$ Å for PBO). For the assumed rodlike conformation, $x_\eta \simeq x_w(x_z/x_w)^{1/2}$ and $x_s = x_w(x_z/x_w)(x_{z+1}/x_z)^{1/2}$, so that x_z/x_w and x_{z+1}/x_z can be estimated by comparison of x_w, x_η, and x_s. The results, given in Table I suggest a rather broad molecular-weight distribution for the unfractionated polymer, with some reduction in x_z/x_w for the fractions.

The satisfactory picture described above for the data on PBO-6 does not obtain with the other PBO polymers studied or with the PPTA or Kevlar polymers examined. Instead, as shown in Table II, both M_w and $[\eta]$ appear to depend drastically on the solvent used, with $[\eta]$ increasing as M_w decreases. The same trend can be observed on the addition of lithium sulfonate to solutions of PBO, PPTA, or Kevlar in chlorosulfonic acid, as shown by the data in Tables II and III and Figure 2. For the polymers studied, $[\eta]$ can be fitted with the empirical relation

$$[\eta]^{-1} = [\eta]_{CSA}^{-1}[1 + KI^{1/2}] \qquad (5)$$

TABLE II

Polymer	Solvent[a]	$10^{-4}M_w$	$10^{12}\langle s^2 \rangle_{LS}$	δ	$[\eta]$
PBO-9	CSA	0.25	31	1.0	7.4
PBO-30		0.27	42	1.0	5.0
PPTA-2		0.21	43	1.0	4.65
Kevlar		0.69	58	1.0	22
PPTA-2	CSA +0.01 N LiCAS	1.27	15	—	2.24
Kevlar		2.52	34	—	11.0
PBO-2	MSA	2.2	51	0.5	2.55
PBO-6		0.2	32	1.0	1.00
PBO-9		2.33	21	0.4	3.80
PPTA-2		1.67	13	0.3	1.93
Kevlar		4.53	25	0.3	8.83
PPTA-2	96% H_2SO_4	2.07	19	0.4	1.30
Kevlar		5.78	16	0.3	6.53

[a] CSA = chlorosulfonic acid; LiCSA = lithium chlorosulfonate; MSA = methane sulfonic acid.

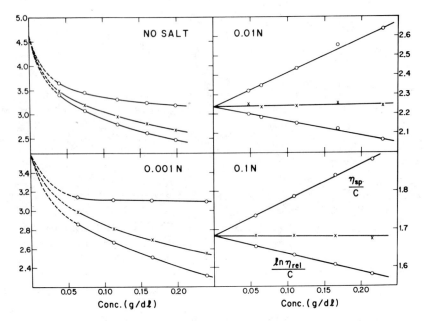

FIG. 2. Viscometric data on a PPTA polymer in chlorosulfonic acid containing lithium chlorosulfonate as indicated. The crosses are calculated with the one-point intrinsic viscosity formula $[\eta] = \{2(\eta_{sp} - \ln \eta_{rel})/c^2\}^{1/2}$.

where $[\eta]_{CSA}$ is the intrinsic viscosity in chlorosulfonic acid and the ionic strength I of the solution includes not only the added salt but also the counterions produced by the protonation of the polymer, assuming one proton per amide or oxazole residue. The dashed curves in Figure 2 were constructed with the aid of eq. (5), similar results obtaining with all three polymers studied.

The dependence of the observed values for $[\eta]$ and M_w on the concentration of added salt for the high-molecular polymers suggests that the ionic strength of the solvent may be an important parameter in understanding the results reported in Table II. Since conduction in sulfonic acids occurs by a proton transfer

TABLE III
Effect of Solvent in the Intrinsic Viscosity

	Polymer			
Solvent[a]	PBO-9	PBO-30 (Intrinsic Viscosity, dl/g)	PPTA-2	Kevlar
CSA	8.5	6.00	4.65	22
CSA + 0.001 N LiCSA	—	—	3.19	17.0
CSA + 0.01 N LiCSA	—	3.64	2.24	11.0
CSA + 0.1 N LiCSA	4.06	3.36	1.68	8.60
CSA + 1.0 N LiCSA	—	—	1.63	—
MSA	3.80	3.16	1.93	8.83
H_2SO_4 + 4 M H_2O	—	—	1.34	6.53

[a] CSA = chlorosulfonic acid; LiCSA = lithium chlorosulfonate; MSA = methane sulfonic acid.

mechanism, an estimate of their ionic strength can be made using the observed specific conductivity κ_{sp}. The latter is given in Table IV, along with some other physical data on several sulfonic acids of interest here. All these acids are believed to undergo self-protonation to generate ions according to the reaction [10, 14]

$$2RSO_3H \rightleftharpoons RSO_3H_2^+ + RSO_3^-$$

In some cases, ions can also be produced by acid-base reactions of RSO_3H with the products RH and SO_3 of dissociation of the sulfonic acid. In addition, of course, the polymers of interest here act as weak bases toward RSO_3H, becoming polycations and creating RSO_3^- anions as counterions:

$$P + nRSO_3H = PH_n^+ + nRSO_3^-$$

where P represents the polymer. For the system of interest here, κ_{sp} can be expressed in the form [10]

$$\kappa_{sp} = 10^{-3} \sum_i \lambda_i C_i \tag{6}$$

where λ_i is the ion conductance of species i present in molar concentration C_i. If only self-protonation need be considered in the conductance of pure chlorosulfonic acid, the ionic strength I is found to be $10^3 \kappa_{sp}/(\lambda_{RSO_3H_2^+} + \lambda_{RSO_3^-})$. Since for sulfonic acids, it is reported [10] that $\lambda_{RSO_3H_2^+}/\lambda_{RSO_3^-} = 1.45$, the relation for the ionic strength reduces to $I = 10^3 \kappa_{sp}/2.45\lambda_{RSO_3^-}$ or $I = 2 \times 10^{-3}M$ for chlorosulfonic acid if the value of $\lambda_{ClSO_3^-}$ given in Table IV is used. Although this ionic strength is low, it should be sufficient to suppress the so-called "polyelectrolyte effect" that can occur when the Debye length b^{-1} is very large [16]. Typically, this effect leads to large values of $[\eta]$ and low values of M_w from light scattering if I is less than about $10^{-4}M$. For the solvents used here, b^{-1} calculated from the usual relation

$$b^2 = \frac{8\pi N_A e^2}{1000\epsilon kT} I \tag{7}$$

TABLE IV
Physical Data for Some Sulfonic Acids (RSO$_3$H)[a]
(25° Unless Noted Otherwise)

R	Density ρ (g/cc)	Melting temperature T_m (°C)	Dielectric constant ϵ (cgs units)	Viscosity η (cP)	Specific conductivity $10^4 \kappa$ (ohm^{-1} cm^{-1})	Ion conductance (ohm^{-1} cm^2) $\lambda_{RSO_3H_2^+}$	$\lambda_{RSO_3^-}$
HO-	1.827	10.4	100	24.5	103.2	220	150
CH$_3$-	1.481	20	—	10.5	41.8	—	75
Cl-	1.753	−80	(60)[b]	2.43	4.0[c]	—	70
F-	1.726	−89	120	1.56	1.09	—	135

[a] Data from references 10–14.
[b] Value cited as uncertain in original reference.
[c] Reported value for fresh, undistilled commercial grade acid.

is about 45 A° in chlorosulfonic acid, as compared with 17 A° for methane sulfonic acid, using the data in Table IV. Consequently, it seems probable that even in chlorosulfonic acid the ionic strength is great enough to suppress the polyelectrolyte effect in the determination of $[\eta]$ and M_w. On the other hand, the ionic strength varies widely, over the range 0.002–0.01 for the sulfonic acids used here, so that the effects on $[\eta]$ and M_w observed with different sulfonic acids are in qualitative accord with those obtained with added salt.

Data on $[\eta]$ and M_w as a function of the elution volume V_e suggest a possible model to explain the unexpected variation of $[\eta]$ and M_w with solvent. In Figure 3 data on fractions of PPTA-2 in methane sulfonic acid are compared with similar results on a heterocyclic polymer BBL, described in detail elsewhere [1]. The data with $[\eta]$ and M_w determined directly on the PPTA solutions removed from the chromatograph agree well with the correlation between $[\eta]M_w$ and V_e reported for BBL, whereas the values of $[\eta]M_w$ are lower than expected for the PPTA fractions characterized on redissolution following recovery by precipitation and drying. This disparity is more marked in Figure 4, giving $[\eta]$ as a function of V_e. A model in which a low degree of intermolecular aggregation obtains with the rodlike chains associated in parallel arrays, without appreciable effect on the average length L, can rationalize these observations. Under these circumstances, the aggregation reduces the number N of effective chains by the factor ν_w of chains per bundle (weight average) without much effect on the viscosity increment per bundle, resulting in a decreased viscosity for the solution;

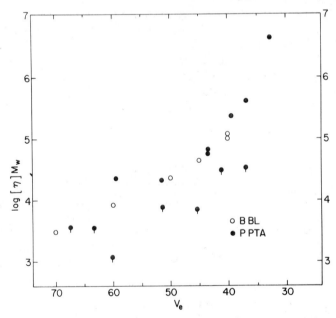

FIG. 3. The product of molecular weight M_w and intrinsic viscosity versus elution volume for fractions of PPTA-2 and BBL. Filled circles: data determined directly in the solutions removed from the chromatograph for PPTA-2; filled circles with pips: data determined on the redissolved PPTA-2 following recovery by precipitation and drying.

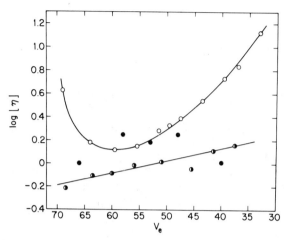

FIG. 4. Intrinsic viscosity $[\eta]$ versus elution volume V_e for PPTA-2. Open circles: direct determination; half-filled circles: redissolution of precipitate.

that is, according to eq. (3) the viscosity increment per number N of effective chains per unit volume is

$$[\eta]M/N_A = \frac{\pi}{100} L^3 f(L/d, L/\rho) \qquad (8)$$

where $[\eta]$ and M are the measured values, including the effects of association if present. For the prescribed association, the only effect on $[\eta]M$ is through the small effect on d. For example, for rodlike chains this factor would vary as $\nu_w{}^{0.2}$ with the type of association postulated, making $[\eta]M$ nearly insensitive to association for small ν_w. Qualitatively, this effect may explain the observed tendency of $[\eta]$ to decrease and M_w to increase as the ionic strength is increased. As I is increased, the Debye length becomes so short that the electrostatic interactions needed to keep the rodlike chains from associating are effectively shielded, permitting the weak attractive interchain interactions to hold a few chains together in a more or less parallel array. Accordingly, the smallest measured M_w and the largest $[\eta]$ are those most nearly representative of a true solution, meaning the data in chlorosulfonic acid, the solvent of lowest ionic strength used here, for the systems studied here. Of course, this does not guarantee a true solution at all molecular weights even with chlorosulfonic acid. It may be noted that the unusual dependence of η_{sp}/c and $\ln \eta_{rel}/c$ on c given in Figure 2 for data in chlorosulfonic acid can be understood as a consequence of this association, with the apparent value of $[\eta]$ decreasing as the counterion concentration increases with the addition of polymer. In addition, the agreement between data on $[\eta]M_w$ versus V_e for the (never precipitated) fractions of PPTA-2 and the data on BBL probably does not indicate that the former were true solutions but merely that the degree of association was low enough so that the simple model described above could be applied. Evidently, this is not the case with the precipitated PPTA-2 fractions, since $[\eta]M_w$ is depressed for these materials.

If, despite indications to the contrary, it is assumed that data on $[\eta]$ and M_w for solutions of the (never precipitated) fractions of PPTA-2 in methane sulfonic acid are free of the effects of association, the data can be analyzed with eq. (3) to determine ρ and d, using the Kuhn approximation to evaluate M_L. The results, shown in Figure 5, give $\rho = 100$ Å and d 6 Å using the function $f(L/d, L/\rho)$ for a wormlike cylinder according to Yamakawa and Fujii [7]. Similar results obtain if this analysis is applied to data on fractions of PPTA reported by Arpin et al. [16] and unfractionated PPTA reported by Schaefgen et al. [17]. In the analysis given here, we take M_w to be a reasonable molecular-weight average for correlation with $[\eta]$, since the latter is about proportional to molecular weight for the PPTA fractions studied. Although the value of ρ found for PPTA is larger than commonly observed with vinyl polymers, it only encompasses about eight repeating units of the chain, so that PPTA cannot be considered truly rodlike in dilute solution if the data on solutions in methane sulfonic acid are accepted as reliable and free from effects of interchain association. In our opinion this conclusion is fallacious, owing to the effects of intermolecular association. Instead, the data on solutions of PPTA in chlorosulfonic acid indicate a larger persistence length, probably around 500 Å.

The data on $[\eta]$, $\langle s^2 \rangle_{LS}$, and M_w for the unfractionated PPTA-2, Kevlar, and PBO polymers 9 and 30 in chlorosulfonic acid deviate from the results expected on the basis of the data on PBO-6 and its fractions. For example, estimates of x_z/x_w based on x_η are unexpectedly large, and estimates of x_{z+1}/x_z are unacceptably small (less than unity). These results may be the consequence of residual association obtaining even in solutions in chlorosulfonic acid or might reflect the polyelectrolyte effect discussed above. We hope to distinguish between these alternatives in future work.

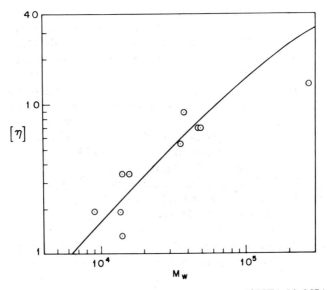

FIG. 5. The intrinsic viscosity versus molecular weight for fractions of PPTA-2 in MSA. The curve represents eq. (3) calculated with values of M_L, ρ, and d given in the text.

Studies of Concentrated Solutions

The supramolecular structure discussed above with the chains in parallel arrays is made even more favorable as the polymer concentration is increased into the range of moderately concentrated solutions (a few percent up to about 10%). At high enough concentration, the polymers studied here form optically anisotropic solutions and exhibit the phenomenon of flow opalescence. In the following we discuss some rheological data on both isotropic and anisotropic solutions of PBO and Kevlar.

For many materials, the dependence of the steady-state viscosity η_κ and recoverable compliance R_κ on the shear rate κ can usefully be represented in terms of a time constant $\tau_c = \eta_0 R_0$ [18]

$$\eta_\kappa = \eta_0 Q(\tau_c \kappa) \tag{9}$$

$$R_\kappa = R_0 P(\tau_c \kappa) \tag{10}$$

For example, significant departure of $Q(\tau_c \kappa)$ from unity usually occurs when κ is about τ_c^{-1} (the parameter $\tau_c \kappa$ is an example of a class of dimensionless parameters termed "Deborah numbers" by Reiner [19]. Markovitz has shown that the use of the time constant τ_c in eq. (9) is consistent with the time-temperature superposition principle of linear viscoelasticity for a class of materials termed thermorheologically simple [20].

Examples of eqs. (9) and (10) can be illustrated with theoretical relations for small $\tau_c \kappa$, calculated with neglect of intermolecular effects, with the results of Kirkwood and Plock [21] for η_κ and Kotaka [22] for R_0 for rods,

$$\eta_\kappa = \eta_0[1 - 1.2952(\tau_c \kappa)^2 + 3.1940(\tau_c \kappa)^4 + \cdots] \tag{11}$$

and with the results of Bird, Evans, and Warner for rigid dumbbells [23],

$$\eta_\kappa = \eta_0[1 - 1.4286(\tau_c \kappa)^2 + 5.3151(\tau_c \kappa)^4 + \cdots] \tag{12}$$

$$R_\kappa = R_0[1 - 1.1791(\tau_c \kappa)^2 + \cdots] \tag{13}$$

The numerical results of Kirkwood and Plock can be approximated over a wider range of $\tau_c \kappa$ by the empirical relation

$$\eta_\kappa = \eta_0/[1 + 0.626\tau_c|\kappa|] \tag{14}$$

which fits the numerical results for isolated rods within 7% for $\tau_c \kappa < 3$.

The data in Figure 6 illustrate the application of eqs. (9) and (10) to results obtained with a solution of Kevlar (2% polymer in MSA). The dashed curve in Figure 6 corresponds to data on a solution of polyisobutylene in cetane [4]. It can be seen that nonlinear effects are observed at comparable values of $\tau_c \kappa$, with the flexible polymer polyisobutylene and the more rodlike Kevlar, illustrating the usefulness of the parameter τ_c as a criterion for the onset of nonlinearity independent of chain conformation. Conversely, the nature of the reduced function η_κ/η_0 versus $\tau_c \kappa$ does not provide information on chain conformation, at least not over the range of $\tau_c \kappa$ studied here. Similar data for optically isotropic solutions of rodlike PBO are illustrated in Figure 7.

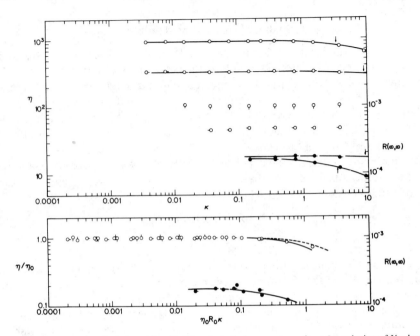

FIG. 6. Log η_κ (open circles) and log R_κ (closed circles) versus log κ for a 2% solution of Kevlar in MSA (upper figure) and the reduced plot according to eqs. (9) and (10) (lower figure). Temperature key: O, 8.75°; O-, 25.7°; O, 49°; Q, 72°.

The rheological behavior observed with optically anisotropic solutions is significantly different from that given in Figures 6 and 7 for isotropic solutions. Some results obtained with a solution of Kevlar in methane sulfonic acid with a polymer concentration high enough ($w_2 = 0.09$) for the solution to exhibit optical anistropy are given in Figure 8. Although the recovery could be measured with an optically isotropic solution of Kevlar just slightly lower in concentration ($w_2 = 0.081$), the behavior obtained with the anisotropic Kevlar solution after cessation of steady state flow did not permit an estimate of R_0. In the latter case the recovery was very small, leading to an unreasonably small value of R_0. Apparently, the optically anisotropic solution acts as a weak solid after cessation of steady state flow, trapping the molecular structure in the nonrandom array characteristic of the steady-state flow. This postulate is consistent with the viscometric behavior at small κ exhibited in Figure 8, in which the viscosity, defined as the shear stress divided by the shear rate, increases, apparently without bound, as κ decreases. In an effort to compare the data on Kevlar solutions with concentrations of 0.02, 0.081, and 0.09, we have constructed the plot of η_κ/η_{REF} versus $\tau_c\kappa$ given in Figure 9, where η_{REF} is equal to η_0 for the two optically isotropic Kevlar solutions but is taken to be the viscosity in the "plateau region," e.g., $0.01 < \kappa < 0.03$, for the data at 8.8° for optically anisotropic solution. Similarly, τ_c is equal to $\eta_0 R_0$ for the two isotropic solutions, but τ_c is arbitrarily adjusted to force the data on the anisotropic solution to coincide with the data on the two isotropic solutions for $\tau_c\kappa > 0.1$. This treatment is motivated by the possibility that supramolecular structure obtaining in the quiescent solution or

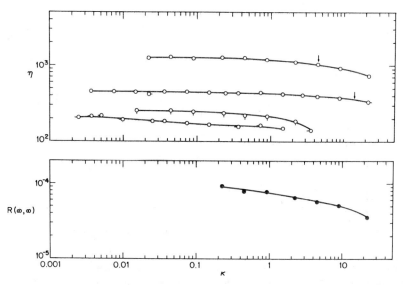

FIG. 7. Log η_κ (open circles) and log R_κ (closed circles) versus log κ for a solution with 6.05 wt % PBO-2 in MSA. O, 4.9°; O-, 239°; O, 45.6°; -O, 53.6°. The arrows indicate the shear rate where $\eta_0 R_{0\kappa} = 1/2$.

even at low κ is disrupted at larger κ, making it reasonable to expect comparable behavior at large enough $\tau_c\kappa$. The results, shown in Figure 9, give η_{REF} equal to 3.2×10^5 and 2.2×10^4 at 8.8 and 52°, respectively, and R_0 calculated as τ_c/η_{REF} equal to 6×10^{-5} cm²/dyn. The latter compares with R_0 equal to 1.8 $\times 10^{-4}$ and 2.6×10^{-5} cm²/dyn for the isotropic solutions with w_2 equal to 0.02

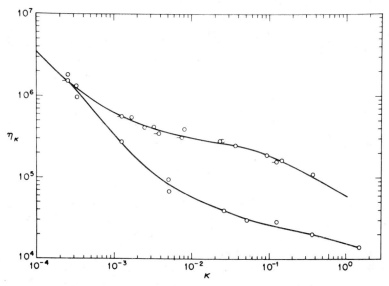

FIG. 8. Log η_κ versus log κ for a solution of Kevlar in MSA ($W_2 = 0.09$) at two temperatures: O, 52.2°; -O, 8.8°.

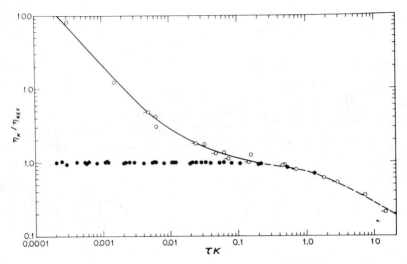

FIG. 9. Log η_κ/η_{REF} versus log $\tau_{c\kappa}$ for the data from Figure 8 and data on Kevlar solutions with $w = 0.02$ and 0.08 (both shown as O). The latter were plotted with $\eta_{REF} = \eta_0$ and $\tau_c = \eta_0 R_0$.

and 0.081, respectively, suggesting that R_0 actually increases somewhat as optical anisotropy is developed with Kevlar solutions.

The results of similar but more systematic experiments with PBO-2 are given in Figure 10. Optically anisotropic solutions were obtained at 25° for w_2 greater than about 0.08. The viscometric data are similar to results reported by Hermans and others for poly(benzyl-L-glutamate) (PBLG) and to recent data on PPTA, Kevlar, and poly(p-benzamide) (PBA) [24–28]. With this polymer, values of R_0 entered in Figure 10 could be measured directly from the recovery after steady-state flow for the data shown, but, as with Kevlar, it was not possible to determine R_0 for solutions with concentrations much greater than the concentration w_2^* for the onset of optical anisotropy. The tendency for R_0 to decrease with increasing w_2 for isotropic solutions and for R_0 to increase with increasing w_2 for w_2 near w_2^* noted with Kevlar is also found with PBO-2.

It is of interest to compute the value (L/d) corresponding to the concentration w_2^* for the onset of the separation of the order anisotropic phase. For this purpose, we use the results of Flory's theory for the phase separation of an assembly of noninteracting rods, leading to the relation [29]

$$\phi_c = \frac{8}{y_n} \left(1 - \frac{2}{y_n} \right) \tag{15}$$

for the minimum volume fraction ϕ_c of polymer required for noninteracting chains with number-average length to diameter ratio y_n to develop a stable ordered phase. The result gives $L_n/d = 85$ for PBO-2. It is difficult to compare this with the value of M_w entered in Table II, owing to the unknown molecular-weight heterogeneity of the specimen and the uncertain effects of association on M_w. Assuming d to be 5 Å, the cited value of L_n/d gives $M_n = 8100$, with $M_L = 19.15$ daltons/Å, in comparison with $M_w = 2.2 \times 10^4$ entered in Table II for

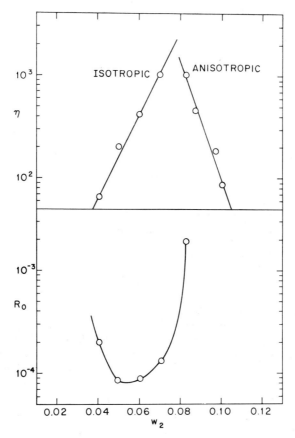

FIG. 10. Viscosity η_{REF} and recoverable compliance R_0 versus weight fraction W_2 for PBO-2 in MSA at 25°.

PBO-2 in methane sulfonic acid. The estimate for M_n appears surprisingly high in comparison with the observed M_w, perhaps reflecting the inadequacy of the assumption that the rodlike PBO chains can be treated as noninteracting rods.

The temperature dependence of η_0, or η_{REF}, observed with a solution of PBO-2 in methane sulfonic acid ($w_2 = 0.0825$) is given in Figure 11. The difference between the curves for decreasing and increasing temperature probably is due to the introduction of a small amount of moisture into the sample during the experiment. The solutions are optically isotropic at temperatures greater than T, the temperature for the maximum in the viscosity, and are completely anistropic for temperatures less than T_2, the temperature for the minimum in the viscosity. At temperatures between T_2 and T_1 the solution contains both isotropic and anisotropic regions, determined by observation of the solution on a microscope stage between crossed polars. The behavior found with PBO is similar to that reported for PPTA and PBA. A schematic diagram indicating the combined effects of temperature and composition for polymers such as PBO, PPTA, and PBA is given in Figure 12.

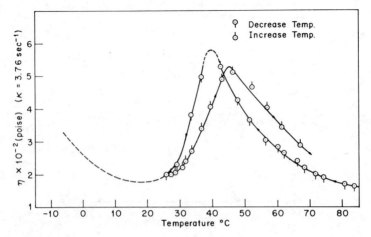

FIG. 11. Steady-state viscosity η at hear rate $\kappa = 3.76$ sec^{-1} versus temperature for a PBO-2–MSA solution ($w_2 = 0.0825$).

The time constant τ_c is an important variable to consider in the characterization of the rheological properties of solutions of rodlike polymers. For the data in Figure 10, τ_c increases monotonically with increasing c, exhibiting a sharp increase as the solution becomes anisotropic, as shown in Figure 13. One important practical consequence of this is that nonlinear behavior, which is significant for $\tau_c\kappa > 1$, is encountered at unusually low values of κ for the anisotropic solution. Another consequence is increased duration of order induced in high shear flow ($\tau_c\kappa > 1$) on cessation of flow. The time constant τ_c provides a useful measure of the latter time scale, but perhaps a better indication is given by the

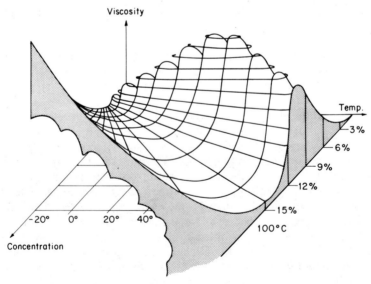

FIG. 12. Schematic diagram for the combined effects of temperature and composition for the rodlike polymer solutions.

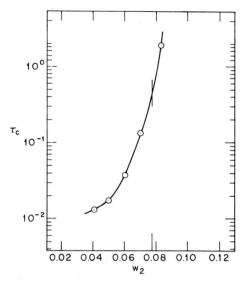

FIG. 13. Dependence of τ_c calculated as $\eta_0 R_0$ on the weight fraction w_2 of polymer for PBO-2 in MSA. The vertical bar indicates where the viscosity is maximum by extrapolation from Figure 10.

time $\theta_{1/2}$ required for the recovery $\gamma_{R\kappa}$ to reach one-half of its final value after cessation of steady state flow. Some data for $\theta_{1/2}$ as a function of the rate of shear γ for the steady flow prior to recovery are given in Figure 14 for an anisotropic solution of PBO-2 in methane sulfonic acid (8.25 wt %, 41°). For this solution, τ_c was found to be 0.04 sec. It can be seen that $\theta_{1/2}$ decreases somewhat with increasing κ but is about 100-fold larger than τ_c. Evidently, one would have sufficient time to quench the flow induced orientation, say, by removal of solvent, before appreciable disorientations could take place.

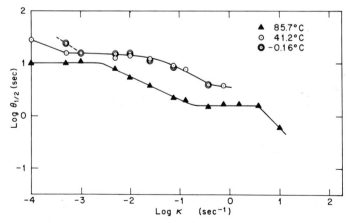

FIG. 14. Time $\theta_{1/2}$ for half-recovery versus shear rate κ for PBO-2–MSA ($w_2 = 0.0825$) at three temperatures.

Comments on Fiber Spinning of Rodlike Polymers

One of the objectives of our studies on PBO and PPTA is to elucidate some of the factors that are important in the fabrication of fibers and films with a high degree or order with the rodlike chains in parallel array over large distances. For example, fiber spinning of PPTA from optically anisotropic concentrated solutions is known to give fibers with a high "instantaneous" tensile modulus E_0. Presumably, the high value of E_0 arises from a high degree of long-range intermolecular order, and the latter is induced by the flow orientation of the rodlike chains. The molecular orientation in elongational flow with rate of elongation κ can be calculated for a simplified model following the procedure of Bird et al. [30] using Kramer's method [31] for an irrotational flow (to be discussed elsewhere). Since the important effects of intermolecular interaction are not included in this calculation, the results can have only qualitative significance for concentrated solutions of rodlike polymers. The molecular orientation in elongational flow can conveniently be represented with an angular distribution function $\psi(\theta)$, where $\psi(\theta) \sin \theta \, d\theta$ is proportional to the fraction of rodlike chains that have their axis at an angle θ with the flow direction. The normalized distribution $\psi(\theta)/\psi_0$ is shown as a function of θ in Figure 15 for several values of $\tau_c \bar{\kappa}$. It can be seen that appreciable orientation can be achieved for elongation rates large enough so that $\tau_c \bar{\kappa} > 6$. Although we cannot expect this result to be directly applicable with anisotropic solutions with long-range molecular order, perhaps one can use the general result that $\tau_c \bar{\kappa}$ is a useful parameter to the characterized flow conditions necessary to achieve molecular orientation with rodlike chains and that it is necessary for $\tau_c \bar{\kappa}$ to be greater than about 5 to obtain such orientation. According to our results, one advantage accrued through the use of anisotropic solutions is the large values of τ_c that can be obtained without a parallel large increase in the viscosity. In fact, for concentrations just a little greater than the critical concentration necessary to achieve phase separation of the ordered state, it appears that τ_c increases whereas

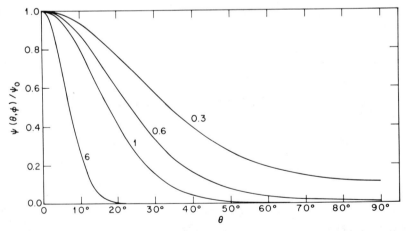

FIG. 15. Relative angular distribution of dumbbells for different values of $\tau \kappa$ for elongational flow.

$(\eta)_{\tau_{cK}=1}$ decreases with increasing concentration. Not only does this have a beneficial effect in the fiber spinning operation itself, but the increased values of τ_c also indicate that disorientation will be slow, allowing time to remove solvents, etc., so that the ordered state can be effectively quenched in the fiber.

With the systems under study here, it appears that it may be difficult to obtain concentrated solutions free of molecular aggregates capable of high orientation, owing to the (metastable) configuration of the aggregate species. Thus, there may be a balance between the advantage in increasing the concentration to increase τ_c and decrease η and the need to maintain a solution free of aggregates of chains in nonparallel arrays. It appears that with PPTA and PBA there may be some advantage in post-fiber-forming heat treatment in which a small permanent elongation is induced at an elevated temperature. Perhaps this post treatment would be necessary if aggregate-free, ordered solutions could be prepared.

It is a pleasure to acknowledge partial support for the work received from the National Science Foundation, Division of Materials Research (Grant No. DMR74-14953), and the Air Force Materials Laboratory, Polymers Branch (Contract No. F33615-76-C-5178). The assistance of D. B. Cotts in the polymerization of the PPTA polymers is also acknowledged.

REFERENCES

[1] G. C. Berry, *J. Polym. Sci. Polym. Symp., 34,* 143 (1978).

[2] T. E. Helminiak and G. C. Berry, *J. Polym. Sci. Symp., 34,* 107 (1978).

[3] T. E. Bair, P. W. Morgan and F. L. Killian, *Macromolecules,* in press.

[4] G. C. Berry and C. P. Wong, *J. Polym. Sci. Polym. Phys. Ed., 13,* 1761 (1975).

[5] C.-P. Wong and G. C. Berry, in "Structure—Solubility Relations in Polymers," F. W. Harris and R. B. Seymour, Eds., Academic, New York, 1977, p. 71.

[6] H. Yamakawa, "Modern Theory of Polymer Solutions," Harper & Row, New York, 1971, p. 56.

[7] H. Yamakawa and M. Fujii, *Macromolecules, 7,* 128 (1974).

[8] H. Nagai, *Polym. J., 3,* 67 (1972).

[9] G. C. Berry, in "Contemporary Topics in Polymer Science," Vol. 2, E. M. Pearce and J. R. Schaefgen, Eds., Plenum Press, 1977, p. 55.

[10] R. H. Flowers, R. J. Gillespie, E. A. Robinson, and C. Solomons, *J. Chem. Soc.,* 4327 (1960).

[11] D. P. Sidebottom and M. Spiro, *J. Phys. Chem., 79,* 943 (1975).

[12] R. C. Paul, K. K. Paul, and K. C. Malholia, *J. Chem. Soc. A,* 2712 (1970).

[13] E. A. Robinson and J. A. Ciruna, *Can. J. Chem., 46,* 1719 (1968).

[14] J. Barr, R. J. Gillespie, and K. C. Thompson, *Inorg. Chem., 3,* 1149 (1964).

[15] H. Eisenberg, "Biological Macromolecules and Polyelectrolytes in Solution," Oxford University Press, London, 1976, p. 88.

[16] M. Arpin and C. Strazielle, *C. R. Acad. Sci. Ser. C, 280,* 1293 (1975).

[17] J. R. Schaefgen, V. S. Foldi, F. M. Logullo, V. H. Good, L. W. Gubich, and F. L. Killian, *Polym. Prepr. Am. Chem. Soc. Div. Polym. Chem., 17*(1), 69 (1976).

[18] G. C. Berry, B. L. Hager, and C.-P Wong, *Macromolecules, 10,* 361 (1977).

[19] M. Reiner, *Phys. Today, 17*(1), 62 (1964).

[20] H. Markovitz, *J. Polym. Sci. Polym. Symp., 50,* 431 (1975).

[21] J. G. Kirkwood and R. J. Block, *J. Chem. Phys., 24,* 665 (1965).

[22] T. Kotaka, *J. Chem. Phys., 30,* 1556 (1959).

[23] R. B. Bird, D. C. Evans, and H. R. Warner, *Adv. Polym. Sci., 8,* 1 (1971).

[24] J. J. Hermans, Jr., *J. Colloid Sci., 17*, 638 (1962).
[25] V. N. Tsvetkov, K. A. Andrionov, Ye. N. Ryumtser, I. N. Shtennikova, N. N. Makarova, and N. A. Kurasheva, *Polym. J. USSR, 15*, 455 (1973).
[26] T. S. Sokolova, S. G. Yefimova, A. V. Volokhina, G. I. Kudryavtsen, and S. P. Papkov, *Polym. Sci. USSR, 15*, 2832 (1973).
[27] S. L. Kwolek, P. W. Morgan, J. R. Schaefgen, and L. W. Gubrich, *Polym. Prepr. Am. Chem. Soc. Div. Polym. Chem., 17*(1), 53 (1976).
[28] S. P. Papkov, V. G. Kulichikhin, V. C. Kalmykova, and A. Ya. Malkin, *J. Polym. Sci. Polym. Phys. Ed., 12*, 1753 (1974).
[29] P. J. Flory, *Proc. R. Soc. London, A234*, 73 (1956).
[30] R. B. Bird, M. W. Johnson, and C. F. Curtis, *J. Chem. Phys., 51* (7), 3023 (1969).
[31] H. A. Kramers, *Physica, 10*, 777 (1943).

RHEOLOGY OF CONCENTRATED SOLUTIONS OF POLY(γ-BENZYL-GLUTAMATE)

GABOR KISS and ROGER S. PORTER

Materials Research Laboratory, Polymer Science and Engineering Department, University of Massachusetts, Amherst, Massachusetts 01003

SYNOPSIS

Steady shear viscosity, dynamic viscosity, dynamic modulus, and normal force were measured via rotational rheometry for concentrated solutions of racemic mixtures of poly(benzyl-L-glutamate) and poly(benzyl-D-glutamate) in *m*-cresol. A transition from the isotropic state to liquid–crystalline order with increase in concentration was indicated by optical anisotropy and maxima in all four material functions. This occurred at a critical concentration higher than the Flory prediction. Over a well-defined range of concentrations and shear stresses, some of the liquid–crystalline solutions exhibited negative first normal-stress differences that were not due to inertial effects.

INTRODUCTION

It has been known for some time that solutions of rigid rodlike particles form anisotropic, i.e., liquid–crystalline, solutions if the concentration is sufficiently high. Well-established theories [1–6] predict that as the concentration of a dilute (isotropic) solution is increased, a phase transition occurs to an anisotropic phase. As concentration is increased further, the fraction of anisotropic phase increases at the expense of the isotropic phase without, however, changing the concentrations in each phase until the solution becomes fully liquid–crystalline. This phenomenon has been demonstrated for systems of a variety of rodlike particles [7–10].

One class of rodlike molecules that has received considerable attention is the synthetic polypeptide poly(γ-benzyl-glutamate), usually the L enantiomer (PBLG). This polypeptide is obtainable in narrow molecular weight distributions and in appropriate solvents forms a helix that behaves in solution like a rigid rod [11]. When a single enantiomer is dissolved in an appropriate solvent, the helices all have the same optical rotary sense, resulting in a cholesteric liquid crystal. However, for equimolar mixtures of both L and D enantiomers, a nematic liquid crystal is obtained [12]. The early developments in this field have been reviewed [13]. The rheology of concentrated solutions of this polymer was studied by Hermans [14] and later by Iizuka [15]. Hermans studied several molecular weights of BPLG in *m*-cresol by capillary flow measurements. Iizuka used solutions of PBLG in CH_2Br_2 and in dioxane, and made steady-shear and oscillatory measurements using a cone-and-plate rheometer.

Journal of Polymer Science: Polymer Symposium 65, 193–211 (1978)
© 1978 John Wiley & Sons, Inc.   0360-8905/78/0065-0193$01.00

The present work is an extension of previous studies with distinctions in the following respects: (1) an equimolar mixture of PBLG and PBDG was used rather than a single enantiomer; (2) a wider range of homogeneous shear rates was accessible through the use of a variety of cones with different radii and cone angles for steady shear in the cone-plate geometry; (3) measurements of dynamic viscoelastic properties were made using the "eccentric rotating disk" geometry. This permits decoupling the viscous and elastic response of the system without the need for mechanical oscillation and the necessity of making precise phase angle measurements. This technique has been shown to give results equivalent to those from oscillatory shear for a variety of polymer solutions and melts [16].

EXPERIMENTAL

PBLG of molecular weight 350,000 and low dispersity was obtained from Biopolymer Corp., of Moreland Hills, Ohio, and PBDG of molecular weight 320,000 and low dispersity was obtained from Pilot Chemical Corp., of Watertown, Massachusetts. All solutions contained equal weights of both enantiomers; this will henceforth be indicated by the acronym PBG.

The (helicogenic) solvent selected was m-cresol in order to minimize concentration changes during measurement due to solvent evaporation. The m-cresol was distilled prior to use. Solutions of 3–10 wt % were made by successive evaporation of a dilute solution in a vacuum evaporator at 80°. Solutions of 25–11 wt % were made by successive dilutions of a 25 wt % solution that was obtained by introducing weighed amounts of PBG and m-cresol into a sealed container. A homogeneous solution was obtained in about 2 weeks.

Because of the high cost of the PBG and the high concentrations required, each solution was reused several times in the rheological measurements. Importantly, it has been shown that shear stresses higher than those achieved in this study neither disrupt the helical structure of PBG of comparable molecular weight dissolved in m-cresol nor cause mechanochemical degradation [17]. The solutions were observed to darken with time, but this was presumed to be due to oxidation of the solvent, which normally darkens on standing. In fact the ease with which m-cresol oxidizes may be a desirable property for inhibiting solute oxidation. After rheological measurements were completed, aliquots of each solution were retained, and the polymer was recovered by dissolving in methylene chloride followed by precipitation with methanol. The recovered PBG was then dissolved in dichloroacetic acid (a nonhelicogenic solvent) to 0.0467 ± 0.00002 wt %. Flow times through a Ubelhode viscometer were compared with that for a solution of fresh polymer. In all cases the flow times were found to be identical, indicating that no degradation had occurred.

Steady-shear viscosity and total normal-thrust measurements were made on a Rheometrics RMS-7200 mechanical spectrometer. Shear rates in the range 0.25–10,000 sec^{-1} were accessible through the use of cone-and-plate geometries of the following radii (in cm) and cone angles (in rad) respectively: 5.00, 0.04; 2.50, 0.04; 2.50, 0.1; 1.25, 0.1; 1.25, 0.01. All data reported herein represented averages of many measurements. Precision in runs was about 5% for both viscosity and normal force; however, run-to-run reproducibility was no better than 20% in some cases. This scatter was not due to instrument instability but to some unknown phenomenon, occurring, possibly, at the platen surface.

An effort was made to obtain data at each shear rate using as wide a variation in geometry as possible. The data at intermediate shear rates were measured over the overlapping ranges for at least two different cones, with data at extremely low and high shear rates measured with one cone only. Data at shear rates above 1000 sec^{-1} do not refer to steady-state measurements due to exudation of sample from the gap at longer times. Reported values were extrapolated to time of initiation of shear. All data were taken with both clockwise and counterclockwise rotation, which gave identical results.

Dynamic viscosities and moduli were measured in the same instrument using the eccentric rotating-disk mode over a rotation speed range of 0.01–62.5 rad/sec. Each reported point represents an average of many measurements generally made with plates of different radii and plate separation. To test whether results were in the linear viscoelastic region, each measurement was made at five different strains ranging from 0.15 to 0.75 at low rotation speeds and 0.03 to 0.15 at high rotation speeds. This procedure reduced the rather large scatter of data, which amounted to about 30% of the mean values reported.

All measurements were made at an ambient temperature of 24° ± 0.5°. Shear heating as measured by a thermocouple embedded in one platen was negligible.

RESULTS

Birefringence

Examination of the PBG solutions in m-cresol by polarized light microscopy indicated that solutions 8.1 wt % polymer and below were optically isotropic and solutions 9.9% and above were optically anisotropic. This result may be compared with conditions expected to produce anisotropy according to theory. For example, PBG of molecular weight 335,000 corresponds to 1506 residues. Since the diameter of the α-helix is 15 Å and the length of the helix is 1.5 Å per residue, the axial ratio would be 150 [18]. The equation given by Flory [3]

$$\phi_2^* = \frac{8}{p}\left(1 - \frac{2}{p}\right)$$

yields a critical volume fraction for formation of an anisotropic phase of 0.053. This corresponds to a concentration of 6.9 wt % for PBG in m-cresol. Thus the Flory theory appears to underestimate slightly the critical concentration for formation of the anisotropic phase. An error in this direction would be expected if the helices were not perfectly rigid [6].

Steady-Shear and Dynamic Viscosities

Steady-shear viscosity measurements were made over a wide range of shear rates for all concentrations. In all cases a low-shear limiting viscosity was obtained, but in no case could a high-shear limit be observed. Plotting the low-shear limiting viscosity reveals a rapid increase with concentration to a maximum at 11.0 wt % followed by an equally rapid decrease to a minimum at 22 wt % (see Fig. 1). At yet higher concentrations there is the suggestion of further gradual increase in viscosity. The shoulder that appears on the low concentration side of the viscosity maximum is unlikely to be due to experimental error. Moreover a comparable feature has been reported by Iizuka [15].

Steady-shear viscosity measurements for the three lowest concentrations did not extend to a shear rate high enough to obtain an unambiguous power law index, (i.e., the exponent n such that $\tau_{12} \propto \dot{\gamma}^n$), but an estimate of $n = 0.12$ can be made (see Fig. 2). The solutions of concentration 9.9 wt % and above all gave measurements in the power-law region; in some cases log η versus log $\dot{\gamma}$ was linear for up to four decades of shear rate (see Fig. 3). The power-law indices obtained for these solutions were in the range $0.45 \le n \le 0.5$.

It can be seen in Figures 2 and 3 that the flow curves for the different solutions intersect, implying that the concentration at which the viscosity reaches a maximum depends on the shear rate. Indeed, it would be expected that the effect of shear orientation would be to ease the formation of the anisotropic phase. This behavior is seen explicitly in Figure 4, which shows that the concentration of maximum viscosity decreases with shear rate, in agreement with the observation of Hermans [14].

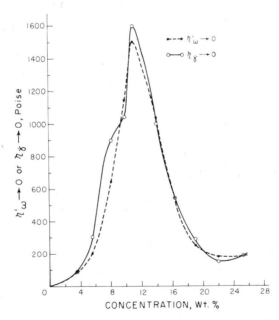

FIG. 1. Low-shear limit of steady-shear viscosity and low-frequency limit of dynamic viscosity versus concentration for PBG in m-cresol.

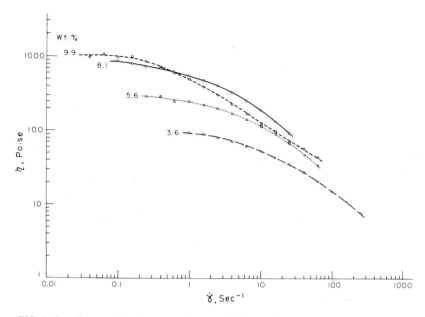

FIG. 2. Steady-shear viscosity versus shear rate for isotropic solutions of PBG in m-cresol.

Dynamic viscosity measurements were made for all concentrations over a frequency range of three decades in most cases (see Figs. 5 and 6). Clear low-frequency limiting values were indicated which, in most cases, agree well with the low-shear limit of η (see Fig. 1). The maximum value of $\eta'_{\omega \to 0}$ was observed at the same concentration as the maximum value of $\eta_{\dot{\gamma} \to 0}$. However, the shoulder on the low concentration side of the viscosity maximum was clearly absent. Also it is not clear whether the dynamic viscosity would have increased gradually with further concentration or had reached a limiting value.

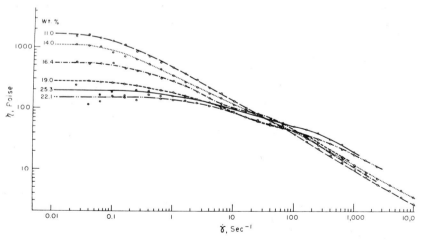

FIG. 3. Steady-shear viscosity versus shear rate for liquid–crystalline solutions of PBG in m-cresol.

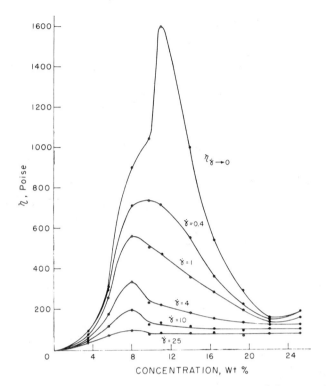

FIG. 4. Concentration dependence of steady-shear viscosity at several shear rates for PBG in *m*-cresol.

The curves of log η' versus log ω for various concentrations (see Figs. 5 and 6) generally did not cross, indicating that the peak in η' versus C would not be shifted to lower concentration by an increase in frequency. Thus, unlike steady shear, dynamic shear does not appear to drive the thermodynamic transition to the anisotropic phase to a lower concentration.

It is also suggested by Figure 6 that plots of log η' versus log ω converge to a common curve at high ω for all concentrations above that at which the viscosity peak was observed. This behavior is in contrast to that of the same solutions in steady shear for which curves of log η versus log $\dot{\gamma}$ all intersect at approximately the same point (see Fig. 3). It is of note that the curves of log η versus log $\dot{\gamma}$ change order at this intersection point; i.e., viscosity decreases with concentration for solutions between 11.0 and 25.3 wt % at lower shear rates and increases with concentration at higher shear rates.

The plateaus in curves of log η' versus log ω and to a lesser extent in log η versus log $\dot{\gamma}$ were experimentally significant and reproducible. They may have been a consequence of the use of a mixture of two polymers (PBLG and PBDG) of slightly different molecular weight.

It is worthwhile to contrast the behavior of solutions on either side of the viscosity-concentration maximum with increasing deformation rate, i.e., shear rate or frequency. The steady-shear viscosities of the isotropic solutions—those

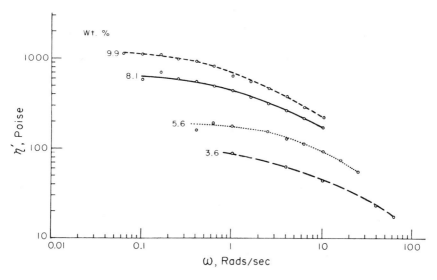

FIG. 5. Dynamic viscosity versus frequency for isotropic solutions of PBG in *m*-cresol.

below the viscosity maximum—was much more dependent on shear rate than those of the liquid–crystalline solutions—those above the viscosity maximum—as indicated by the difference in power-law exponents (see Figs. 2 and 3). The dependence of dynamic viscosity on frequency was also slightly greater for isotropic solutions, though this difference is less evident because of the curvature of the graphs and the restricted range of measurements (Figs. 5 and 6). A more easily seen difference in the dynamic viscosity behavior is that log η' versus log ω tends to converge to a common curve for the liquid–crystalline solutions (Fig. 6) but not for the isotropic solutions (Fig. 5).

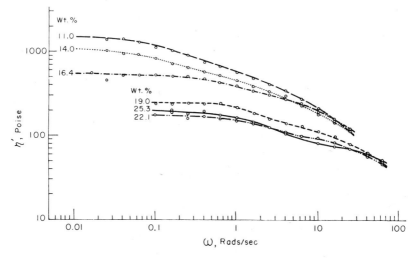

FIG. 6. Dynamic viscosity versus frequency for liquid–crystalline solutions of PBG in *m*-cresol.

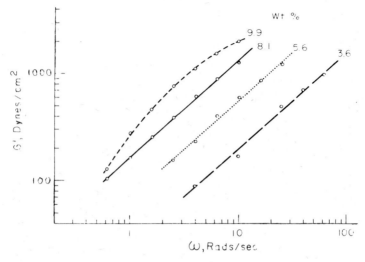

FIG. 7. Dynamic modulus versus frequency for isotropic solutions of PBG in *m*-cresol.

Dynamic modulus

Dynamic modulus G' measurements were made for all solutions over a range of about two decades of frequency ω (Figs. 7 and 8). Log G' versus log ω increases linearly with a slope of slightly less than unity for most concentrations. However, log G' versus log ω for the 9.9 wt % solution is curved and intersects the curve

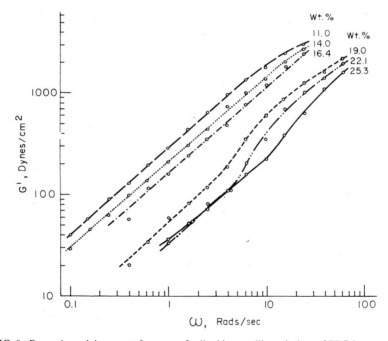

FIG. 8. Dynamic modulus versus frequency for liquid-crystalline solutions of PBG in *m*-cresol.

for 11.0 wt % at a frequency of 1 rad/sec. This means that for frequencies greater than 1 rad/sec the maximum modulus occurs at 9.9 wt %, whereas for frequencies less than 1 rad/sec the maximum modulus occurs at 11.0 wt %. In Figure 9, G' at a frequency of 4 rad/sec is plotted against concentration. It is quite similar in appearance to $\eta_{\dot\gamma\to 0}$ versus C and $\eta'_{\omega\to 0}$ versus C (Fig. 1). There is a suggestion of a shoulder on the high concentration side of the maximum, possibly due to the coexistence of an isotropic phase and an anisotropic phase at these concentrations. This would correspond to a concentration between Robinson's A point and B point [10]. This curve of G' versus C differs from those obtained by Iizuka [15] for PELG in dioxane or CH_2Br_2 in that G' does not appear to increase sharply again at the highest concentrations, where the solutions were fully liquid–crystalline.

Referring again to Figure 8, we observe that the curve of log G versus log ω deviates from linearity for the three highest concentrations at higher frequencies. The onset of these deviations appear to be at frequencies such that G' was approximately constant at 100–200 dyn/cm².

It is significant that the slopes of log G' versus log ω are similar for both liquid–crystalline and isotropic solutions, whereas log η versus log $\dot\gamma$ and log η versus log ω are quite different (see Figs. 2, 3, 5, and 6). The value of this slope, slightly less than unity, is in contrast to the Kirkwood-Auer [19] prediction of a slope of 2 at low frequencies, increasing to an asymptotic value at high frequency. Of course, the K-A theory was derived for dilute solutions of rigid rods and cannot be applied to concentrated solutions. Predictions of the dynamic elastic behavior of liquid–crystalline or concentrated solutions of rigid rods are lacking.

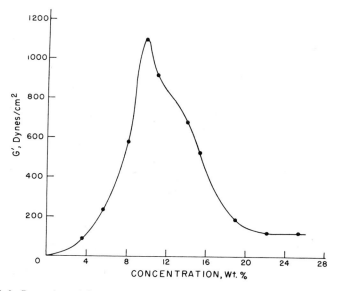

FIG. 9. Dynamic modulus at $\omega = 4$ rad/sec versus concentration of PBG in m-cresol.

First Normal-Stress Difference

The first normal-stress difference, $\tau_{11} - \tau_{22}$, was measured via total normal thrust for flow in cone-and-plate geometry for all solutions over a wide range of shear rates.

The four least concentrated solutions were investigated prior to improvements in instrumentation that greatly extended the accessible range of shear rate. All results are shown in Figure 10. The plots of log $\tau_{11} - \tau_{22}$ versus log $\dot{\gamma}$ are convex and parallel. The maximum value of $\tau_{11} - \tau_{22}$ was exhibited by the 8.1 wt % solution over the entire range of shear rate. Log $(\tau_{11} - \tau_{22})$ versus log $\dot{\gamma}$ for the 11.0 wt % solution is shown in Figure 11. This curve has a peculiar appearance, with an inflection point at 1 sec^{-1} followed by a convex region, another inflection point at 70 sec^{-1}, and finally a long linear region with slope 0.6. It is possible, but not likely, that the four least concentrated solutions would have behaved similarly if a greater range of shear rate had been accessible and that those measurements which could be made fell fortuitously in the convex region.

The curve for $\tau_{11} - \tau_{22}$ for the 14.0 wt % solution is presented in Figure 12. Its appearance is peculiar indeed, with $\tau_{11} - \tau_{22}$ increasing with $\dot{\gamma}$ for low shear rate, reaching a maximum at $\dot{\gamma} \approx 0.2$ sec^{-1}, abruptly becoming negative at $\dot{\gamma} \approx 0.6$ sec^{-1}, reaching a maximum negative value at $\dot{\gamma} \approx 3$ sec^{-1}, and abruptly becoming positive again at $\dot{\gamma} \approx 5$ sec^{-1}, then increasing to a linear region with slope of approximately 0.7. This remarkable behavior was also observed for solutions of concentrations 16.4, 19.0, and 22.1 wt % (see Figs. 13–15). The solution of concentration 25.3 wt % did not exhibit unusual (negative) normal stress behavior (see Fig. 16).

It can be observed in Figures 12–15 that the shear rates at which the reversals in sign of $\tau_{11} - \tau_{22}$ occurred were concentration-dependent. The shear stress

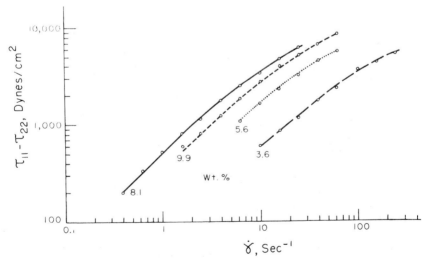

FIG. 10. First normal-stress difference versus shear rate for isotropic solutions of PBG in *m*-cresol.

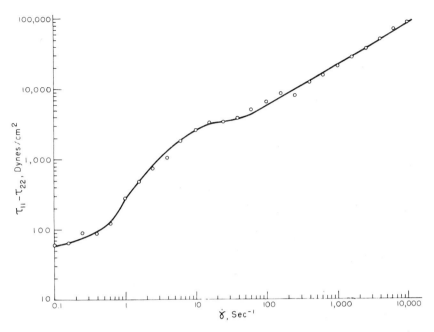

FIG. 11. First normal-stress difference versus shear rate for PBG in m-cresol (11.0 wt %).

to which the fluids were being subjected at these reversal points is shown in Figure 17 as a function of concentration. The experimental points frame an area in which negative first normal-stress differences were observed. The dashed extrapolations intersect at a concentration of 11 wt %, consistent with the positive $\tau_{11} - \tau_{22}$ measured at all shear rates for the 11.0 wt % solution.

An effort was made to exclude the possibility that these unusual (negative) normal stress observations were the result of artifact. It has been known for some time that inertial forces, which are neglected in the conventional analysis relating first normal-stress difference to total normal thrust in cone-and-plate flow, can make a negative contribution to the normal thrust [20]. The inertial contribution can in fact produce a spurious sign reversal in $\tau_{11} - \tau_{22}$ for certain polymer solutions that is eliminated when a correction is applied (see Fig. 18). This correction was applied to all normal-thrust data and found to be much smaller than the measured normal thrust in all our observations of PBG in m-cresol. Since the only material property that influences the magnitude of the inertial correction is density, inertial effects cannot be responsible for the negative first normal-stress difference observed here.

Another precaution that was taken to preclude artifacts was the use of a variety of cones of different radii and cone angles. It would be expected that any contribution due to edge effects or secondary flow would be of very different magnitudes for different cones and would have been immediately obvious.

Finally it should be noted that both total normal thrust, from which $\tau_{11} - \tau_{22}$ is calculated, and torque, from which η is calculated, were measured simultaneously, both on the bottom plate below the rotating cone. Therefore, it would

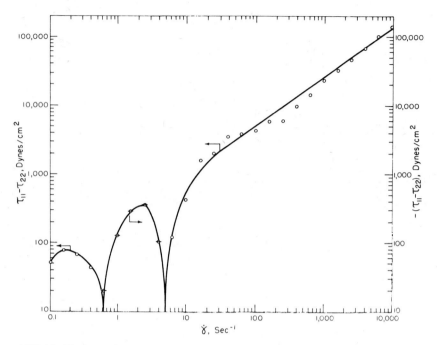

FIG. 12. First normal-stress difference versus shear rate for PBG in m-cresol (14.0 wt %).

be expected that any gross departure from the simple laminar flow field assumed by conventional analyses of cone-and-plate flow would also have been manifested in the apparent viscosity behavior as a discontinuity or change in slope.

A possible rationalization of a negative normal thrust can be found in postulating a dilatometric effect, i.e., a reduction of sample volume on shearing. Efforts to correlate the shape of the free surface at the edge of the cone with negative normal thrust observations were frustrated by changes in shape caused by incipient secondary flow or the fluid's being forced out by centrifugal force. Another pertinent observation is that normal thrust values, both positive and negative, were established almost immediately on initiation of shear and were time-independent for the duration of the measurement (up to 15 min). The same normal thrust values were obtained for both clockwise and counterclockwise rotation. Decay of the normal thrust on cessation of shear was very rapid. All these observations argue against the negative normal thrust's being a dilatometric effect.

DISCUSSION

In the course of the exposition of the experimental results we have had occasion to use the terms *liquid–crystalline* and *isotropic* to describe solutions of different concentrations. Strictly speaking, the term *liquid–crystalline* should be applied to any fluid that exhibits long-range order, e.g., a solution of rodlike molecules at the A point, at which the anisotropic phase first emerges. Previous workers

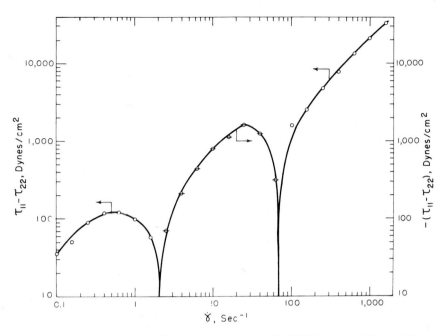

FIG. 13. First normal-stress difference versus shear rate for PBG in *m*-cresol (16.4 wt %).

have identified the concentration of maximum viscosity with the A point [21] and have accordingly labeled solutions of higher concentrations as liquid-crystalline and those of lower concentration as isotropic.

It is unlikely, however, that the formation of a small fraction of an anisotropic

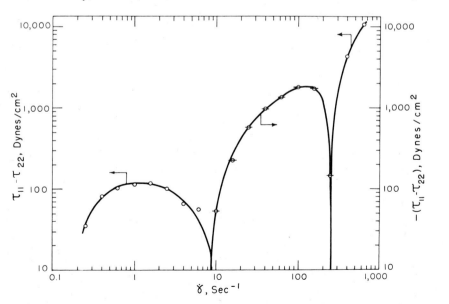

FIG. 14. First normal-stress difference versus shear rate for PBG in *m*-cresol (19.0 wt %).

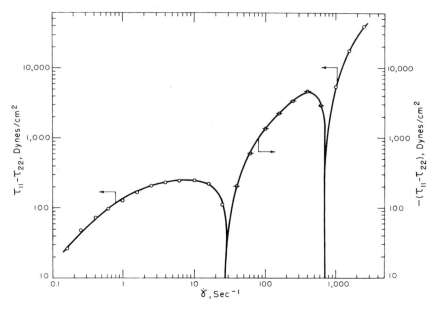

FIG. 15. First normal-stress difference versus shear rate for PBG in m-cresol (22.1 wt %).

phase at the A point would be reflected in an immediate and drastic change in the macroscopic properties of the solution as a whole. It is more likely that the anisotropic phase must comprise a substantial fraction of the solution; this would correspond to a concentration somewhere between the A point and the B point, that at which the solution becomes wholly anisotropic. It would not be unexpected that different material properties of the solution would be sensitive to the presence of an anisotropic phase to varying degrees. Thus depending on which property one were to observe, the solutions would appear to become liquid–crystalline at different concentrations. As shown in Table I, liquid–crystalline order was manifested at different concentrations by various properties, corresponding to different ratios of anisotropic phase to isotropic phase. The Flory prediction of the critical concentration for formation of an anisotropic phase is in fact lower than the indications of any of the observed properties. It is interesting that birefringence, which would be expected to be the most sensitive to the presence of a small amount of anisotropic phase coexisting with a large amount of isotropic phase, gave a higher estimate for the critical concentration than first normal-stress differences.

The steady-shear viscosity measurements presented here are in qualitative agreement with those of previous workers [14, 15] in that viscosity increased sharply with concentration to a maximum value after which it dropped sharply and also in that solutions beyond the viscosity maximum were partially or fully liquid–crystalline. We feel, however, that the viscosity peak does not uniquely mark the emergence of the anisotropic phase but that some of this phase is present before the maximum viscosity is achieved. The magnitudes of the viscosities observed were similar to those previously reported, but detailed

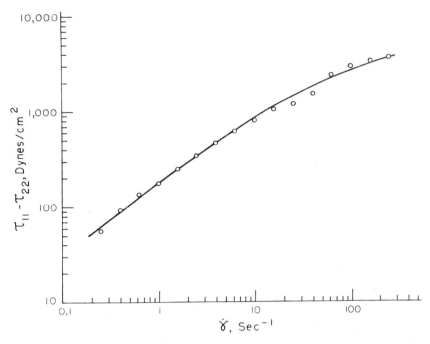

FIG. 16. First normal-stress difference versus shear rate for PBG in m-cresol (25.3 wt. %).

comparisons were impossible, since previous investigators used systems with slight differences, e.g., solvent, molecular weight of PBLG. In particular no previous reports of the rheological behavior of equimolar mixtures of PBLG and PBDG are available

It would be anticipated that concentrated solutions of an equimolar mixture of PBLG and PBDG, which form nematic liquid crystals [10], would exhibit different rheological properties from concentrated solutions of a single enantiomer, which form cholesteric liquid crystals. Cholesteric liquid crystals are generally considered to be special cases of nematics in which adjacent molecular layers are slightly displaced, leading to a helical superstructure with its screw axis perpendicular to the molecular layers [22]. Experiments on low-molecular-weight thermotropic liquid crystals indicate that the viscosity changes at the nematic $\rightarrow$ isotropic-liquid and cholesteric $\rightarrow$ isotropic-liquid transitions are markedly different [23]. Nematics generally have lower viscosities than their

TABLE I

Indication of liquid crystal order	Concentration (wt %)
Optical anisotropy (birefringence)	9.9
Maximum in $\eta_{\dot{\gamma} \rightarrow 0}$ vs c	11.0
Maximum in $\eta_{\omega \rightarrow 0}$ vs c	11.0
Maximum in G' vs c	9.9
Maximum in $\tau_{11} - \tau_{22}$ vs c	8.1
Flory theory	6.9

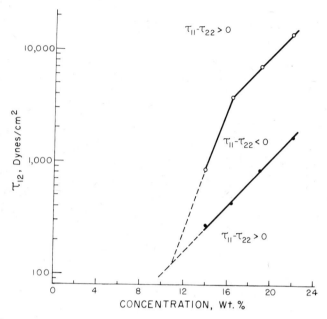

FIG. 17. Shear stress at points of change in sign of first normal-stress difference versus concentration for PBG in *m*-cresol.

isotropic liquids, since their molecules are readily oriented in the direction of flow measurements. Cholesterics generally have higher viscosities than their corresponding isotropic liquids. This behavior may be understood by reference

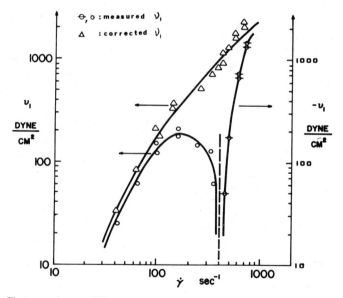

FIG. 18. First normal-stress difference versus shear rate for 2% aqueous polyacrylamide solution ($\overline{M}_n = 1.56 \times 10^6$).

to results obtained by Pochan and Marsh [24], who observed that mixtures of cholesteryl chloride and cholesteryl oleyl carbonate were found to exhibit the Grandjean texture at low shear rate (helical screw axes perpendicular to direction of shear). The effect of increased shear was to tilt the helical screw axes to a direction parellel to the direction of shear (dynamic focal conic texture), meaning that the molecules themselves would be oriented perpendicular to the direction of shear.

Another difference in the rheological behavior of thermotropic cholesterics and nematics is the observation that nematics are Newtonian whereas cholesterics are non-Newtonian [23]. Thus, the presence of cholesteric superstructure has a significant effect on the rheology of thermotropic liquid crystals.

For this reason it is interesting that both nematic and cholesteric solutions of helical polypeptides behave so similarly. The viscosity-concentration behavior of PBLG in m-cresol reported by Hermans [14] is very similar to that of our racemic mixtures and also for solutions of nonchiral rodlike macromolecules [25, 26]. A direct comparison of PELG in CH_2Br_2 with a mixture of PELG and PEDG in the same solvent revealed very similar viscosity behavior [15].

Our observation was that for concentrations up to 25.2 wt %, the liquid–crystalline solutions were of lower viscosity than the isotropic solutions at the highest concentration. It may be argued that since the cholesteric superstructure in solutions of a single enantiomer would be quite weak, probably much weaker than that in thermotropic cholesterics, and easily disrupted by shear, rheological measurements on such solutions were in fact made on nematic liquid crystals. Such a shear-induced cholesteric → nematic transition process would be analogous to the "unwinding" of a cholesteric in a magnetic field [27]. Such an argument would be strengthened by the observation of yield stresses in solutions of a single enantiomer (nominally cholesteric) and not in solutions of mixtures of both enantiomers (nematic).

The observation of normal stresses for liquid crystals has been infrequently reported. Erhardt, Pochan, and Richards [28] have reported normal-stress measurements on mixtures of cholesteryl chloride and cholesteryl oleyl carbonate. They observed behavior consistent with that of a second order fluid, i.e., that $\lim_{\omega \to 0} G'/\omega^2 = \lim_{\dot\gamma \to 0} [(\tau_{11} - \tau_{22})/2\dot\gamma^2]$ [29].

Normal-stress measurements on concentrated solutions of helical polypeptides, including PBLG, were conducted by Iizuka [15, 30]. However, the results were not reported but were used to calculate extinction angles from which the rotary diffusion constant was deduced and thence an apparent particle size for molecular clusters from tables given by Scheraga [31]. Therefore, our report of negative normal forces appears to be unprecedented in the liquid–crystal literature and in fact may be rare in the rheological literature. Coleman and Markovitz [29] demonstrated that for a second-order fluid in slow Couette flow the viscoelastic contribution to normal thrust must have a sign opposite to the inertial contribution on thermodynamic grounds. Walters [32] reports that measurements of first normal stress difference have invariably led to a positive quantity except for one which was later found to be in error. Adams and Lodge [33] reported the possible observation of a negative value of $\tau_{11} - \tau_{22}$ for solutions of polyisobutylene in decalin. This result was obtained by a combination of $\tau_{11} + \tau_{22} - 2\tau_{33}$, obtained from radial variation in normal stress in a cone-and-plate, with measurements of $\tau_{22} - \tau_{33}$ that were of uncertain accuracy. However, in a discussion of this point, Adams and Lodge concluded that negative values of $\tau_{11} - \tau_{22}$ would be curious but not impossible.

Other observations of negative normal stress have been reported. In a study of rheological properties of aqueous lecithin solutions Duke and Chapoy [34] observed a positive normal thrust in steady-shear cone-and-plate flow, but, on reversal of the direction of rotation, the normal thrust became negative and then increased to the positive steady-state value. This effect was attributed to incomplete normal-stress relaxation on cessation of flow.

Time-dependent, negative, normal thrust observations were reported by Huang [35] on melts of an SBS copolymer. In this case a negative normal thrust was generated on initiation of shear but decayed to zero after about 75 sec. This effect was thought to be due to a small volume decrease caused by shearing.

In a personal communication Iizuka [36] reported negative normal stress in solutions of PBLG in CH_2Br_2 of concentration 10 vol % or greater. This is in agreement with our observation of negative normal thrust only in liquid–crystalline solutions of PBG in m-cresol. This effect was ascribed by Iizuka to the adhesive force of the solution.

Our observations differ from those referred to above in that the negative normal thrust was not time-dependent, nor did it appear only on sudden reversal of the direction of rotation.

CONCLUSION

A theory for the rheological properties of liquid–crystalline solutions of rigid rodlike molecules is lacking. Progress has been made in the flow properties of thermotropic (bulk) liquid crystals from a continuum approach [37], and dilute solution theories of rigid rodlike molecule are very successful [19, 38]. However, neither of these could be expected to apply to this work.

Therefore, we have merely described our observations in the thought that they will stimulate progress toward an understanding of, and perhaps a theory for, the rheology of such fluids.

SYMBOLS AND ABBREVIATIONS

C	Concentration, wt %
G'	Dynamic modulus
n	Power-law index
P	Axial ratio
PBDG	Poly(γ-benzyl-D-glutamate)
PBLG	Poly(γ-benzyl-L-glutamate)
PBG	Equimolar mixture of BPLG and PBDG
$\dot{\gamma}$	Shear rate
η	Steady-shear viscosity

$\eta_{\dot\gamma\to 0}$	Low shear limit of steady-shear viscosity
η'	Dynamic viscosity
$\eta'_{\omega\to 0}$	Low-frequency limit of dynamic viscosity
ν_1	First normal-stress difference
τ_{ij}	Stress
$\tau_{11} - \tau_{22}$	First normal-stress difference
ϕ_2^*	Critical volume fraction of solute for phase separation

REFERENCES

[1] L. Onsager, *Ann. N.Y. Acad. Sci.*, *51*, 627 (1949).

[2] A. Ishihara, *J. Chem. Phys.*, *19*, 1142 (1951).

[3] P. J. Flory, *Proc. R. Soc. London*, *A234*, 60 (1956).

[4] E. A. DiMarzio, *J. Chem. Phys.*, *35*, 658 (1961).

[5] J. P. Straley, *Mol. Cryst. Liq. Cryst.*, *22*, 333 (1973).

[6] S. Ya. Frenkel, *J. Polym. Sci.*, *C44*, 49 (1974).

[7] F. C. Bawden and N. W. Pirie, *Proc. R. Soc. London*, *B123*, 274 (1937).

[8] J. D. Bernal and I. Frankuchen, *J. Gen. Physiol.*, *25*, 111 (1941).

[9] S. Ya. Frenkel, V. G. Baranov, and T. I. Volkov, *J. Polym. Sci.*, *C16*, 1655 (1967).

[10] C. Robinson, J. C. Ward, and R. B. Beevers, *Discuss. Faraday Soc.*, *25*, 29 (1958).

[11] P. Doty, J. H. Bradbury, and A. M. Holtzer, *J.A.C.S.*, *78*, 947 (1956).

[12] C. Robinson and J. C. Ward, *Nature*, *180*, 1183 (1957).

[13] R. S. Porter and J. F. Johnson, "Rheology: Theory and Applications," Vol. IV, F. R. Eirich, Ed., Academic, New York, pp. 317-345.

[14] J. Hermans, Jr., *J. Colloid Sci.*, *17*, 638 (1962).

[15] E. Iizuka, *Mol. Cryst. Liq. Cryst.*, *25*, 287 (1974).

[16] C. W. Macosko and W. M. David, *Rheol. Acta*, *13*, 814 (1974).

[17] J. T. Yang, *J.A.C.S.*, *81*, 3902 (1959).

[18] Y. Layec and C. Wolff, *Rheol. Acta*, *13*, 696 (1974).

[19] J. G. Kirkwood and P. L. Auer, *J. Chem Phys.*, *19*, 281 (1951)

[20] W. M. Kulicke, G. Kiss, and R. S. Porter, *Rheol. Acta*, *16*, 568 (1977).

[21] E. T. Samulski and A. V. Tobolsky in "Liquid Crystals and Plastic Crystals," Vol. 1, G. W. Gray and P. A. Winsor, Eds., Halsted Press, New York, 1974, pp. 175-199.

[22] F. D. Saeva, *Mol. Cryst. Liq. Cryst.*, *23*, 271 (1973).

[23] R. S. Porter, E. M. Barrall, II, J. F. Johnson, *J. Chem. Phys.*, *45*, 1452 (1966).

[24] J. M. Pochan and D. G. Marsh, *J. Chem. Phys.*, *57*, 1193 (1972).

[25] S. P. Papkov et al., *J. Polym. Sci. 12*, 1753 (1974).

[26] S. Kwolek, "Optically Anisotropic Aromatic Polyamide Dopes," U.S. Patent 3,671,542.

[27] P. G. DeGennes, *Solid State Commun.*, *6*, 163 (1965).

[28] P. F. Erhardt, J. M. Pochan, W. C. Richards, *J. Chem. Phys.*, *57*, 3596 (1972).

[29] B. D. Coleman and H. Markovitz, *J. Appl. Phys.*, *35*, 1 (1964).

[30] E. Eizuka, *J. Phys. Soc. Jpn.*, *35*, 1792 (1973).

[31] H. A. Scheraga, J. T. Edsall, J. D. Gadd, *J. Chem. Phys.*, *19*, 1101 (1951).

[32] K. Walters, "Rheometry," Halsted Press, New York, 1974, p. 88.

[33] N. Adams and A. S. Lodge, *Philos. Trans. R. Soc.*, *256*, 149 (1964).

[34] R. W. Duke and L. L. Chapoy, *Rheol. Acta*, *15*, 548 (1976).

[35] T. A. Huang, Ph.D. Thesis, University of Wisconsin, Department of Engineering Mechanics, 1976, p. 93.

[36] E. Iizuka, personal communications, April 1977.

[37] F. M. Leslie, *Arch. Rat. Mech. Anal.*, *28*, 265 (1968).

[38] M. C. Williams, *A.I.C.E.J.*, *21*, 1 (1975).

ULTRAHIGH-MODULUS FIBERS FROM RIGID AND SEMIRIGID AROMATIC POLYAMIDES

GIAN CARLO ALFONSO, ESTELLA BIANCHI, ALBERTO CIFERRI, SAVERIO RUSSO, FRANCO SALARIS,* and BARBARA VALENTI

Istituto di Chimica Industriale and Centro di Studi Chimico-Fisici di Macromolecole Sintetiche e Naturali, C.N.R., University of Genoa—16132, Genoa, Italy

SYNOPSIS

Intrinsic viscosity, concentrated-solution viscosity, and fiber properties of poly(p-benzamide) (PBA) and the polyterephthalamide of p-aminobenzhydrazide (X-500) are presented. PBA yields high-modulus fibers from solution spinning of anisotropic dopes, and X-500 yields high-modulus fibers from isotropic solutions. Both polymers exhibit a wormlike conformation. However, with a low-molecular-weight PBA (10,500) an essentially rigid conformation prevails. The viscosity of both polymers goes through a maximum and a minimum on increasing polymer concentration. However, for PBA the maximum and minimum are observed at low shear stress, and for X-500 the discontinuity is observed only at high shear rate. The initial modulus of low-molecular-weight PBA fibers is unaffected by LiCl concentration but increases with polymer concentration of anisotropic dopes. X-500 (molecular weight, 41,000) is unable to form an anisotropic solution at rest. The initial modulus of X-500 is unaffected by LiCl but depends slightly on polymer concentration. Both shear and elongational flow contributions appear to be secondary for truly stiff polymers. Orientation is controlled by polymer concentration, and the modulus of low-molecular-weight PBA attains the ultrahigh level even for "as-spun" fibers. On the contrary, control of shear and elongational parameters is shown to be essential, and feasible, in order to match the as-spun modulus of PBA, with semirigid polymers unable to form anisotropic solutions at rest.

INTRODUCTION

Ultrahigh-modulus–high-strength fibers composed of aromatic polyamides have recently been produced. Du Pont Kevlar [1] appears to be based on structures of the poly(p-benzamide) (PBA) or the poly(p-phenyleneterephthalamide) type. Monsanto X-500 [2] is a polyterephthalamide of p-aminobenzhydrazide. Both fibers may exhibit initial moduli in excess of 1000×10^9 dyn/cm^2 (800 g/den). Tensile strengths on the order of 30×10^9 dyn/cm^2 have been reported for the du Pont polymer. Because of the high melting temperature

* Permanent address: EUTECO S.p.A., SEAN—20,100, Paderno Dugnano (Milano), Italy.

Journal of Polymer Science: Polymer Symposium 65, 213–222 (1978)
© 1978 John Wiley & Sons, Inc.   0360-8905/78/0065-0213$01.00

of these polymers, which suggests a high degree of chain rigidity, fibers are processed by solution spinning. Chain rigidity is also reflected in the ability of PBA to form anisotropic solutions at rest [1–6] above a critical value of polymer concentration. In fact, the use of anisotropic dopes is believed to be a key factor in the attainment of the superior mechanical properties of Kevlar [1, 4]. On the other hand, X-500 also exhibits superior mechanical properties in spite of its inability [5] to form anisotropic solutions at rest. Thus, mechanical orientation during and after the spinning process also plays a significant role, and a study of the various variables leading to ultrahigh-modulus properties is needed.

Our work presents a comparative analysis of chain rigidity, rheological behavior, and tensile properties for a rigid (PBA) and a semirigid (X-500) polymer. The aim of the comparison is the elucidation of the basic mechanism that allows a semirigid polymer to develop ultrahigh-modulus properties in spite of the inability to form a nematic phase at rest.

EXPERIMENTAL

Materials

PBA was prepared from p-aminobenzoic acid following the method of Yamazaki et al. [7] as described in more detail elsewhere [6]. The intrinsic viscosity in 96% H_2SO_4 at 25° was 1.54 dl/g, corresponding to $M_w \approx 10,500$ [8, 9]. X-500 was prepared by the method of Preston [2]. The intrinsic viscosity in dimethyl sulfoxide (DMSO) at 25° was 4.82 dl/g, corresponding [10] to $M_w = 41,000$.

Polymer solutions (concentration C_p) were prepared in DMSO and dimethylacetamide (DMAc) and with various concentrations (C_s) of added lithium chloride.

Intrinsic Viscosity

Linear plots of specific viscosity versus polymer concentration allowed a simple extrapolation of the intrinsic viscosity [η]. Measurements were performed as previously described [11].

Rheological Behavior

Steady-shear rate measurements were carried out at 20° using a Weissenberg rheogoniometer (model R.18) equipped with a cone and plate. Shear rates in the range 10^{-1}–10^3 sec^{-1} were applied. Occasionally, with X-500, a Couette cylinder-cylinder assembly was used. No difference in results was observed.

Fiber Spinning

Fibers were obtained from a small-scale wet-spinning apparatus by extruding small quantities of polymer dope (about 10 ml) into a coagulating bath 95 cm long. The apparatus has been described in detail by Dyson and Montgomery [11]. Monofilaments were spun through a 100-μ hole·into water at a volume outflow of 0.02–0.16 ml/min and were collected at the end of the bath by a set of rollers. The shear rate at the spinncrette was between 1000 and 10,000 sec^{-1}. An "air gap" was used only in the case of X-500. The filament continued into a washing bath and after about 1 m in air was wound up by a bobbin. The highest take-up rate used along the spinning line, divided by the linear outflow rate, defines the pull-off ratio. Spools were washed in deionized water and dried at room temperature under vacuum. The fibers had circular cross section with diameters between 20 and 50 μ (about 1–2 mils), constant along the filament ($\pm$5% for PBA, $\pm$10% for X-500). No voids or capillaries were evident by optical microscopy. The homogeneity of the filaments is probably due to the formation of a gel phase, as discussed by Ziabicki [13].

Mechanical Properties

Stress-strain curves were determined with an Instron machine, Model 1122, at a stretching rate of 5 mm/min. Before testing, the fibers were stored for at least 2 days in a desiccator. Each test was performed at room temperature on bundles of four untwisted filaments 50 mm long. The reported figures are averaged results on three bundles; the accuracy is within 10%.

RESULTS AND DISCUSSION

Chain Conformation

The results of Arpin and Strazielle [8] indicate that in 96% H_2SO_4, PBA is a wormlike chain with a persistence length $q \approx 400$ Å, a value similar to that observed for chains such as DNA [11]. A somewhat smaller figure, $q = 240$ Å in 96% H_2SO_4, was reported by Schaefgen et al. [9]. By assuming an average value of $q \approx 320$ Å, the meaning of a persistence length of this size may be appreciated using the parameters of the p-benzamide unit indicated in Table I. The persistence length corresponds to a hypothetical rod having 48 units, a molecular weight M_n about 5700, and an axial ratio x of about 64. Flory's theory [3] predicts the critical polymer volume fraction above which a stable anisotropic solution forms to be $v_2^c \approx 8/x$.

On the basis of the latter relationship, a PBA solution composed of unconnected rodlike units with a length equal to the persistence length should become anisotropic when $v_2 \gtrsim 12\%$. Since the molecular weight M_w of our PBA sample was 10,500 and, as reported by Schaefgen et al. [9], the ratio M_w/M_n is about 2, we have actually worked with a solute conforming to the rigid rod model. The conclusions above are strictly valid for PBA in H_2SO_4. In the case of the

TABLE I
Persistence Length Analysis for PBA and X-500

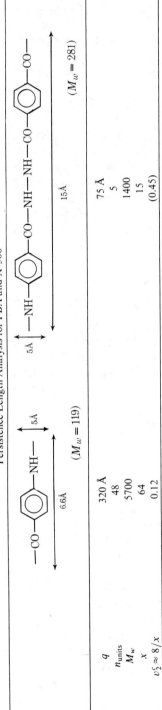

q	320 Å	75 Å
n_{units}	48	5
M_w	5700	1400
x	64	15
$v_2^c \approx 8/x$	0.12	(0.45)

DMAc–LiCl solvent used in the present investigation, an alteration of the persistence length due to LiCl might, in principle, occur. However, the intrinsic viscosity data reported in Figure 1 cannot be readily interpreted in terms of chain conformation. The peculiar temperature behavior, unpublished light-scattering data, and results obtained by Schaefgen et al. [9] indicate the occurrence of considerable aggregation, particularly at high C_s. We note, for example, in Figure 1, the high value of $[\eta]$ at low C_s, which is actually higher than the value in H_2SO_4. Evidently, the dilute solution behavior requires additional investigation.

In the case of X-500, Burke [10] has reported the relation $[\eta] = 6.15 \times 10^{-5} M_v^{1.06}$, which suggests that in this case also we are dealing with a wormlike chain. Our data on intrinsic viscosity versus LiCl concentration in DMSO (Fig. 1) reveal no aggregation and allow a tentative determination of the persistence length on the basis of Ullman's theory [14] of intrinsic viscosity of wormlike chains. In this way, we obtain $q = 75$ Å, which should not, however, be considered a reliable value, in view of the difficult evaluation of the parameters involved. A complete light-scattering analysis is in progress. This result suggests, however, a greater flexibility than for PBA. The meaning of a persistence length of this size may be appreciated considering the parameters of the X-500 unit indicated in Table I. The persistence length corresponds to a hypothetical rod of molecular weight about 1400 and to an axial ratio of 15. Although we cannot strictly apply Flory's theory to a semirigid polymer, by assuming that the solution of X-500 contains rodlike segments equal to the persistence length, we can calculate a limiting value of $v_2^c \approx 45\%$. Since the molecular weight of our sample was 41,000, we have actually worked with a semirigid chain containing several persistent elements arranged in a wormlike manner.

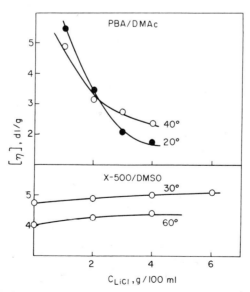

FIG. 1. Variation of the intrinsic viscosity with LiCl concentration for PBA ($M_w = 10,500$) in DMAc and for X-500 ($M_w = 41,000$) in DMSO. Temperature (°C) is indicated.

ALFONSO ET AL.

Rheological Behavior

Typical variations of viscosity with polymer concentration for concentrated solutions of PBA and X-500 are reported in Figure 2. The behavior of PBA is similar to that reported by other investigators [4, 15, 16] who have studied systems exhibiting a transition to a nematic phase on increasing polymer concentration. The viscosity increases at first with C_p, when only the isotropic solution exists, and then decreases when the low-viscosity anisotropic phase appears. When only the anisotropic phase is present, the viscosity increases again with C_p. The maximum and the minimum can be regarded as the limits, at the particular value of shear stress τ, of the narrow biphasic region predicted by Flory [3, 6, 16]. Since η versus τ plots exhibit a plateau between $\tau \approx 5$ and $\tau \approx 100$

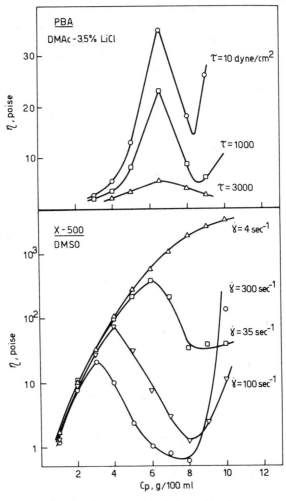

FIG. 2. Variation of the viscosity of concentrated solutions with polymer concentration for PBA in DMAc–3.5 g/100 ml LiCl and for X-500 in DMSO. Shear stress τ and shear rate $\dot{\gamma}$ are indicated. $T = 20°$.

dyn/cm^2 (cf. also Papkov et al. [17]), the critical concentration at $\tau = 10$ dyn/cm^2 (Fig. 2) is close to the extrapolated value at $\tau = 0$. The latter values were used for constructing the phase diagram C_p versus C_s that was presented elsewhere [6]. At high τ, the maximum and the minimum tend to disappear, and the overall viscosity is reduced because of the orienting effect of shear flow.

In the case of X-500 (Fig. 2) marked time effects were observed at low shear rate. For this reason all data are reported at constant shear rate $\dot{\gamma}$. No time effects were noticed for the curve at $\dot{\gamma} = 4$ sec^{-1}, which exhibits no maximum and minimum, indicating that an anisotropic solution at rest is not formed. A similar conclusion was derived from the study of the phase diagram of X-500 previously reported [5]. Evidently, conventional crystallization prevents attainment of a high enough polymer concentration (cf. Table 1) at which an anisotropic phase might be observed. At high $\dot{\gamma}$, η decreases with time and reaches a plateau on the order of 10 min. The plateau values of η are reported in Figure 2. A maximum and a minimum are observed. The discontinuity is shifted to lower C_p on increasing $\dot{\gamma}$. This effect is superficially similar to that observed for PBA, and, in fact, we believe it represents the clue to the development of high-modulus properties for semirigid polymers. Current theoretical descriptions of the non-Newtonian viscosity of concentrated solutions of flexible polymers do not predict such an effect. A more detailed analysis of the observed behavior will be reported later.

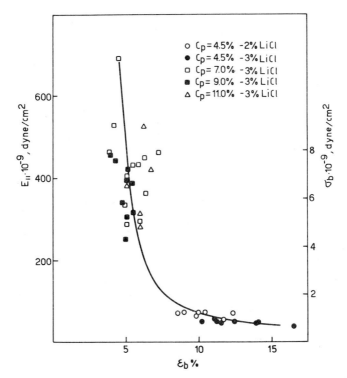

FIG. 3. Initial modulus E_{11}, tensile strength σ_b, and elongation to break ϵ_b for PBA as-spun fibers. The composition of dopes from which fibers were spun is indicated.

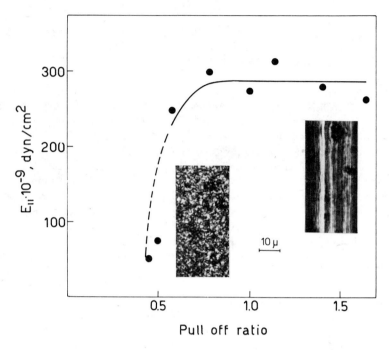

FIG. 4. Initial modulus and optical micrographs of PBA as-spun fibers as a function of pull-off ratio V_1/V_0. V_0 (outflow rate) = 10 m/min. Polymer concentration of the anisotropic dope 9 g/100 ml.

Fiber Properties

PBA fibers were spun from DMAc solutions containing 3 g/100 ml LiCl and C_p = 4.5, 7, 9, 11 g/100 ml and, also, from a solution containing 2 g/100 ml LiCl and C_p = 4.5 g/100 ml. Mechanical properties of these fibers are reported in Figure 3. The polymer concentration clearly influences the modulus, tensile strength, and elongation to break. On the other hand, salt concentration has a negligible effect, indicating that any effect of LiCl on the axial ratio of PBA, resulting either from aggregation or from changes of persistence length, does not affect fiber properties. The large effect of polymer concentration is, of course, similar to that already described by the du Pont investigators [1, 4] and is the basis of their outstanding patent. It will be noticed from Figure 2 that C_p = 4.5 g/100 ml corresponds to an isotropic solution and C_p = 9 and 11 g/100 ml correspond to anisotropic dopes. Flory's theory of rigid-rod equilibria predicts that the disorientation parameter y can be reduced by increasing C_p of the solution at rest. Inclusion of the effect of elongational flow in Flory's theory [18] confirms that an increase of polymer concentration is even more effective in decreasing y than the extensional flow itself. Thus, the increase of E_{11} with C_p observed in Figure 3 is directly related to the decrease of y with C_p. The values of E_{11} reported here are somewhat greater than those reported by the du Pont investigators [1, 4] for as-spun aromatic polyamide fibers. The latter investigators were able to reach the 1300×10^9 dyn/cm^2 level only after heat treatment

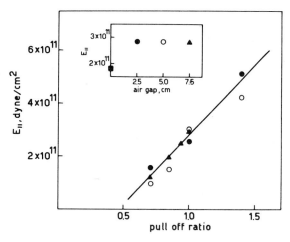

FIG. 5. Initial modulus of X-500 as-spun fibers plotted against pull-off ratio V_f/V_0. $V_0 = 20$ m/min. Insert: modulus versus air gap.

subsequent to fiber formation. The theoretical modulus reported by Fielding-Russell [19] is 1950×10^9 dyn/cm^2. On the other hand, our tensile strengths are smaller than those quoted by the du Pont investigators. Since the molecular length of our polymer is similar to the persistence length of PBA, we were probably able to achieve a better orientation, without special control of elongational flow or drawing, than in the case of longer chains, when the wormlike character is fully evident. The effect of the tension along the spinning line on the as-spun modulus of PBA (Fig. 4) appears to be secondary, provided a small tension is applied during the coagulation of the filament. In fact, optical microscopy reveals that the typical fiber pattern does not appear for pull-offs less than 0.5, corresponding to situations in which the fiber is barely under tension (Fig. 4). It is possible that unless a small tension is applied, anisotropic domains do not become well oriented along the continuous filament.

In a preprint of this work [20] we advanced E_{11} values for as-spun PBA fibers 20–30% higher than those reported here. Fluctuations in the E_{11} value appear to result from different polymer preparations; a more detailed study of the effect of the molecular weight on PBA properties is in progress. Very high-modulus values occasionally observed [20] may be related to accidental drawing during winding.

In the case of X-500, fibers were spun from DMSO solutions containing 0 and 5 g/100 ml LiCl, and $C_p = 10$ and 16 g/100 ml. Stress-strain data for as-spun X-500 indicated that E_{11} is unaffected by C_s, and, at variance with the behavior of PBA, is not greatly affected by polymer concentration. Moduli of the order of 200×10^9 dyn/cm^2 were normally obtained for C_p values greater than those corresponding to the discontinuity on the η versus C_p curves (Fig. 2). Moreover, at variance with PBA, a considerable increase of E_{11} could be obtained (Fig. 5) by using an air gap or increasing the pull-off ratio, which directly affects the elongational flow. The suggestion we wish to draw from these results is that, with a moderate degree of chain rigidity, inadequate to give a nematic phase at rest, it is still possible to produce orientation in the flow field to attain as-spun moduli on the order of 500×10^9 dyn/cm^2. Improvements allowing an approach to the high values of E_{11} observed with nematic solutions appear to be possible. However, these improvements must rely on a careful control of extensional variables, including molecular weight, and on drawing treatment subsequent to fiber formation.

We are greatly indebted to Dr. J. Preston for very useful advice and clarifying discussions.

This investigation was supported by the Italian National Research Council, through its Technological Committee, Research Contract No. 2/76.00081.11.

REFERENCES

[1] S. L. Kwolek, U.S. Patent 3,671,542 to E.I. du Pont de Nemours and Co. (1972).

[2] J. Preston, *Polym. Eng. Sci.*, *15*, 199 (1975).

[3] P. J. Flory, *Proc. R. Soc. London Ser. A*, *234*, 73 (1956).

[4] P. W. Morgan, *Polym. Prepr. Am. Chem. Soc. Div. Polym. Chem.*, *17*(1), 47 (1976).

[5] A. Ciferri, *Polym. Eng. Sci.*, *15*, 191 (1975).

[6] F. Salaris, B. Valenti, G. Costa, and A. Ciferri, *Makromol. Chem.*, *177*, 3073 (1976).

[7] N. Yamazaki, M. Matsumoto, and F. Higashi, *J. Polym. Sci. Polym. Chem. Ed.*, *13*, 1373 (1975).

[8] M. Arpin and C. Strazielle, *Makromol. Chem.*, *177*, 581 (1976).

[9] J. R. Schaefgen, V. S. Foldi, F. M. Logullo, V. H. Good, L. W. Gulrich, and F. L. Killian, *Polym. Prepr. Am. Chem. Soc. Div. Polym. Chem.*, *17*(1), 69 (1976).

[10] J. J. Burke, *J. Macromol. Sci. Chem.*, *A7*(1), 187 (1973).

[11] G. Bressan, R. Rampone, E. Bianchi, and A. Ciferri, *Biopolymers*, *12*, 2227 (1974).

[12] D. E. Montgomery, Ph.D. Thesis, University of Bradford, 1971.

[13] A. Ziabicki, "Physical Fundamentals of Fiber Formation," Wiley, New York, 1976.

[14] R. Ullman, *J. Chem. Phys.*, *49*, 5486 (1968).

[15] J. Hermans, Jr., *J. Colloid Sci.*, *17*, 638 (1962).

[16] W. G. Miller, J. H. Rai, and E. L. Lee, "Liquid Crystals and Ordered Fluids, Vol. II, R. Porter and J. Johnson, Eds., Plenum, New York, 1974, p. 243.

[17] S. P. Papkov, V. G. Kulchikhin, V. D. Kalmykova, and A. Ya. Malkin, *J. Polym. Sci. Polym. Phys. Ed.*, *12*, 1753 (1974).

[18] G. Marrucci and A. Ciferri, *J. Polym. Sci. Polym. Lett. Ed.*, to appear.

[19] G. S. Fielding-Russell, *Text. Res. J.*, *41*, 861 (1971).

[20] G. C. Alfonso, E. Bianchi, A. Ciferri, S. Russo, F. Salaris, and B. Valenti, *Polym. Prepr. Am. Chem. Soc. Div. Polym. Chem.*, *18*, 179 (1977).

Author Index

Alfonso, G. C., 213
Allendoerfer, R. D., 63
Amdur, S., 63
Aoki, H., 29

Beever, W. H., 41
Berry, G. C., 107, 125, 143, 173
Bianchi, E., 213

Casassa, E. F., 125
Chen, C.-Y., 55
Ciferri, A., 213
Coffin, D. R., 29

Day, D., 73
De Gennes, P. G., 85

Ehrlich, P., 63
Enkelmann, V., 73

Fellers, J. F., 29

Hancock, T. A., 29
Harwood, D., 29
Helminiak, T. E., 107
Hofferbert, W. L., Jr., 13

Kiss, G., 193
Kou, L., 91

Lando, J. B., 73
Lenk, R. S., 29

Miller, W. G., 91
Morgan, P. W., 1

Nguyen, H. X., 63

Ohnuma, H., 173

Pecora, R., 79
Piirma, I., 55
Pincus, R., 85
Porter, R. S., 193
Preston, J., 13

Russo, S., 213

Salaris, F., 213
Stille, J. K., 41, 125

Tohyama, K., 91

Valenti, B., 213
Voltaggio, V., 91

White, J. L., 29
Wong, C.-P., 173
Work, J. L., 125

Published Polymer Symposia

All the above symposia can be individually purchased through the Subscription Department, John Wiley & Sons.